Motivating
Mathematics

Engaging Teachers and Engaged Students

Motivating Mathematics

Engaging Teachers and Engaged Students

David Wells

Imperial College Press

Published by

Imperial College Press
57 Shelton Street
Covent Garden
London WC2H 9HE

Distributed by

World Scientific Publishing Co. Pte. Ltd.

5 Toh Tuck Link, Singapore 596224

USA office: 27 Warren Street, Suite 401-402, Hackensack, NJ 07601

UK office: 57 Shelton Street, Covent Garden, London WC2H 9HE

Library of Congress Cataloging-in-Publication Data
Wells, D. G. (David G.)
  Motivating mathematics : engaging teachers and engaged students / David Wells, Fox &
Howard Literary Agency, UK.
    pages cm
  Includes bibliographical references and index.
    ISBN 978-1-78326-752-1 (hardcover : alk. paper) -- ISBN 978-1-78326-753-8 (pbk. : alk. paper)
  1. Mathematics--Study and teaching. 2. Mathematics--Study and teaching (Secondary).
3. Mathematics teachers--Training of. I. Title.
  QA135.6.W445 2015
  510.71--dc23
                             2015021511

British Library Cataloguing-in-Publication Data
A catalogue record for this book is available from the British Library.

Typeset by Stallion Press
Email: enquiries@stallionpress.com

Printed in Singapore

Also by David Wells

Can You Solve These? 1, 2 and 3, Tarquin 1982–1986

Engaging Mathematics I, II and III, WSIHE 1989

Mathematics through Problem Solving, Blackwell 1987

Hidden Connections, Double Meanings, CUP 1988

The Penguin Dictionary of Curious and Interesting Numbers, 1986

The Penguin Dictionary of Curious and Interesting Geometry, 1991

The Penguin Book of Curious and Interesting Mathematics, 1997

The Penguin Book of Curious and Interesting Puzzles, 1992

Problem Solving and Investigations, Rain Press 1986

You are a Mathematician, Penguin 1995

The End of Civilisation, Rain Press 2003

Power and Economics, Rain Press 2001

Prime Numbers, Wiley 2005

Philosophy and Abstract Games, Rain Press 2010

The Third Entity, Rain Press 2012

Economics on the Ropes, Rain Press 2012

Games and Mathematics: Subtle Connections, CUP 2012

Preface: Themes and Sources

This book is identical to *What's the Point? Motivation and the Mathematics Crisis*, published by the author in 2008, apart from the correction of some typos and the deletion of the second Appendix. It is written from the perspective of a one-time teacher who has since written on mathematics education as well as producing a number of 'popular' maths books. Quotations and references have been included from various sources but these should be seen as illustrative only: this is not an academic treatise and at no time did I attempt a comprehensive literature search beyond my initial check on ERIC to see what had been said, or not said, about motivation *vis-à-vis* maths education and my searches among the extensive shelves of the London Institute of Education library.

I have frequently drawn on my own past writings listed in the Author's Bibliography, especially the following which all illustrate my central and continuing themes.

The magazine, *Acid Rain*, later titled *Studies of Meaning, Language and Change*, of which 23 issues were published between 1977 and 1988, included several articles on mathematics education and education in general, among others: 'Education: Ideals, Appreciation and Respect' (1977); 'Problems, Games, Familiarity' (1979a); 'Maths and Morality' (1979b); 'The language of 'problem' and 'investigation'' (1987c); 'General Concepts, and Teaching' (1987d); 'False Simplicity and True Simplicity' (1987e); 'Epistemology of abstract games and mathematics' (1988a); and 'Epistemology of Mathematics, and Teaching' (1988b).

The theme of appreciation was developed further in *Teaching and Appreciation* (1979c).

Between 1980 and 1983, eight issues of *The Problem Solver* were published. It was a magazine of problems for school pupils and their teachers often with newsletters which also contained occasional short articles on problem solving. The premises of *The Problem Solver* were that, 'mathematics is about solving problems, that pupils of all ages and abilities are able to solve problems and that they enjoy doing so'. The problems were presented with no hints or aids and later collected in three small books, *Can You Solve These?* (1982–1984). I agree totally with Paul Halmos that problem solving is 'the heart of mathematics' (Halmos 1980) which perfectly fits the idea, beloved of mathematics educators, that 'mathematics is not a spectator sport'. Problem solving should also be at the heart of pupils' learning.

Three Essays on the Teaching of Mathematics (1982), was a pamphlet of 48 pages which include a tipped-in full-colour reproduction of Dali's *Metamorphosis of Narcissus*, plus actual samples of wallpaper to show their structure and so illustrate the first essay on 'Structure through problem solving'. The other essays were 'Ambiguity and interpretation' and (promoting once again the idea of appreciation) 'Broad concepts and metaphor'. These illustrated my belief that mathematics can be seen as based on perception (as well as being both game-like and scientific) and the claim that pupils need to learn to appreciate mathematics by learning more *about* mathematics.

I later wrote for Blackwell, *Mathematics Through Problem Solving* (1987a;1990a) having previously started (1968–1974: it was not finished) a secondary course based on problem solving for John Murray, two of whose claims were that it gave pupils an overview and an appreciation of mathematics and that 'It does not offer CSE students a watered down O-level course. It does offer the same basic understanding to everyone.' The Blackwell material failed and not entirely because the publishers chose at the very last moment to package it as a giant file of photocopiable loose sheets with a high price to match. A little of the Blackwell material is used in this book, including some references to the 'Introduction' addressed to teachers on themes such as using problems with pupils, changing pupils' and teachers' expectations, building pupils' understanding of the attitudes and expectations of professionals, heuristics and what I call general concepts, the role of time pressure and surreptitious learning and the contrast between true simplicity (rich complexity) and false simplicity.

The original motive for introducing 'investigations' was to offer pupils the chance to think and behave more like actual mathematicians. Unfortunately, 'investigations' in practice rapidly became institutionalised on what I called the data-pattern-generalisation (DPG) model which greatly over-emphasised the scientific side of mathematics at the expense of the game-like and almost entirely excluded proof. *Problem Solving and Investigations* (1986a) followed by a corrected edition (1987) and a third enlarged edition in 1993 sketched the history of the investigations movement and argued just this point and therefore in favour of a much broader conception of mathematical problem solving. 'Investigations and the Learning of Mathematics' (1995a) in *Mathematics Teaching*, also argued against current ideas of investigations.

The title of *Hidden Connections, Double Meanings* (1988c) is more or less explicit, highlighting two sources of motivation which appear again here. The introduction linked mathematics to traditional riddles and humour and individual chapters discussed aspects of perception, skeletons and structures, invariants, metaphors, multiple meanings, the game of algebra and proof.

Two articles, 'Which is the most beautiful?' (1988e) and 'Are these the most beautiful?' (1990b), were based on a quiz originally given at an Easter conference of the Mathematical Association then edited and republished for readers of *Mathematical Intelligencer* who are mostly professionals of one sort or another. Respondents were asked to rank 12 mathematical propositions from one to ten for their 'beauty'. The results are presented in Ch. 5: the most striking was that mathematicians showed a considerable measure of disagreement in their judgements, supporting the conclusion that mathematicians have different styles of doing and thinking about mathematics.

In 1989 I was invited by Afzal Ahmed to work with some A-level students at the West Sussex Institute of Higher Education. This meant some fascinating research including tape recording pupils talking about their attitudes to mathematics, their views on mathematics and beauty and so on. The result was *Engaging Mathematics I, II* and *III*, which are now available as free downloads from the National STEM Centre website.

'Games as a metaphor for mathematics' was the self-explanatory title of a presentation to the Discussion Group on the Philosophy of Mathematics at the British Congress of Mathematics Education (1991a) convened by

Paul Ernest. I still find this analogy very powerful, both educationally and philosophically.

The manuscript of *Ways of Knowing: The Nature, Learning and Teaching of Mathematics* (1992) was privately distributed but never published. Chapter 7 was on 'Affect, motivation and satisfaction'. Many of the arguments in this book are taken more or less directly from that text, usually without attribution. The book was originally titled *Rich Complexity, False Simplicity* so that theme was prominent and it also discussed general concepts and appreciation, the roles of ambiguity and interpretation, and of language, structure and perception, and the game-like and scientific aspects of mathematics.

You Are A Mathematician (1995b), albeit implicitly, presented mathematics as game-like *and* scientific *and* perceptual with chapters on 'Mathematics as science', and 'The games of mathematics', 'Creating new mathematical games', 'Perception and imagination', 'Mathematics in science' and 'The enjoyment of mathematics'. The penultimate chapter presented a 'miniature world' and the final 'chapter' consisted of a long sequence of 'frames' presenting a 'mathematical adventure' that invited the reader to start at the beginning and then at the end of each frame, deciding which frame to go to next. This was intended as a simulation, on paper rather than on screen, of the tactics and strategies of a game-like mathematical exploration.

Mathematics and Abstract Games: an Intimate Connection (2007) republished by Cambridge University Press in 2012 as *Games and Mathematics: Subtle Connections*, is devoted to exploring the similarities (and differences) between mathematics and abstract games such as nine men's Morris, chess and go, so making explicit the three perspectives on mathematics that had appeared less clearly in *You Are A Mathematician* and linking maths and abstract games to the theme of formality in the everyday world, from music and dancing to the theatre and the law courts. The original edition had five appendices which discussed certain philosophical problems about maths and abstract games. Many philosophers and mathematicians have made a connection between mathematics and games only to dismiss it as misleading or insignificant: I believe that it is deep and extremely significant, not least for teachers and pupils. This theme has been analysed further in *Philosophy and Abstract Games* (2010) and *The Third Entity: a Philosophy of Abstract Games* (2012).

*****

I shall tend to refer to problems for which negligible pre-knowledge is required as puzzles, following our everyday usage in which traditional puzzles and those in newspaper columns and popular books have this feature. Problems which require some pre-knowledge to be understood will be called, simply, problems.

References

References to the Author's Bibliography in pages 269–271 will be given with a date and page number, if applicable, only. References with a name followed by a date are to the main bibliography starting on page 273.

Acknowledgements

Passages from the Cockcroft Report, *Mathematics Counts* (Cockroft 1982) and the Professor Adrian Smith report, *Making Mathematics Count* (Smith 2004), are reproduced with the permission of the Controller of HMSO and the Queen's Printer for Scotland.

My correspondence with Jennifer Kano (pages 147; 267–8) is quoted with her permission.

The text and diagrams from 'Cut off too soon, Alas!' (pages 26–7) are quoted from *Mathematics in School* (2008) by permission of the editor.

The quotations taken from *Engaging Mathematics*, Volume III (1989a/1990) (pages 93–4) are reproduced by permission of Professor Afzal Ahmed, one-time director of the Mathematics Centre, University of Chichester and now professor emeritus of mathematics education at Chichester University.

I would also like to thank Alice Oven who first decided that this was a suitable book for Imperial College Press, and Alice, Tom Stottor and Catharina Weijman who saw it through production.

Contents

Preamble: Young on *The Teaching of Mathematics* (1907)

While preparing this book I dipped once again into some old volumes which I keep on my shelves and one, not for the first time, caught my eye: *The Teaching of Mathematics in the Elementary and Secondary School* by J.W.A. Young, which was published more than a century ago in 1907. It is very progressive and child-centred in its analyses and recommendations, which, frankly, prompts the thought, 'why have we not made greater progress in the last 100 years and more?' Here are some of Young's arguments and conclusions with brief comments. I hope readers will find them thought provoking, to say the least.

Young discusses at great length the heuristic method and the Perry Movement or 'laboratory method' as it was called in the USA. The,

> dominating thought of the movements is a fuller consideration of the child mind; a sacrifice of the logical, as hitherto regarded, to the psychological; or, rather, a recognition of the fact that no method of instruction is truly logical which is not psychological, which does not pay attention to the constitution of the child's mind. (Young 1907:91)

Where are the *mathematical laboratories* in our schools, lumbered as they are with targets and exams and more exams and more targets?

Young also discusses mathematics and aesthetics.

Mathematics has beauties of its own — a symmetry and proportion in its results, a lack of superfluity, an exact adaptation of means to ends, which is exceedingly remarkable and to be found elsewhere only in works of the greatest beauty ... The beauties of mathematics — of simplicity, of symmetry, of compactness, of completeness — can and should be exemplified even to young children. When this subject is properly and concretely presented, the mental emotion should be that of enjoyment of beauty, not that of repulsion from the ugly and unpleasant.

(Young 1907:44–45)

Young then quotes Poincaré (1897:82):

Those skilled in mathematics find in it pleasures akin to those which painting and music give. They admire the delicate harmony of numbers and of forms; they marvel when a new discovery opens an unexpected perspective; and is this pleasure not esthetic, even though the senses have no part in it?

Motivation is discussed by Young under the heading of *Interest*:

That the pupil should be kept interested in his work, and that hence the work should be presented in an interesting way, in the *most* interesting way, may be called a pedagogic axiom, yet thinkers of the most diverse sorts accuse the teachers of mathematics of transgressing it grievously.

(Young 1907:91)

Young then quotes a writer on science teaching:

There are only two ways to make a thing stick in the mind of the pupil, one is to repeat the thing without variation until it becomes a habit of mind. This is the method of the algebra teacher. The other is to present the thing with such interest to the learner that his whole being responds to the act of accepting and adopting it, and with such intensity that with only one or possibly two presentations of the thing it is indelibly fixed in the mind of the pupil. This is the necessary method in all rational subjects.

(Young 1907:91–92, note 1, quoting Sage 1903:78)

'This is the method of the algebra teacher!' Surely not! Yet can we totally deny the charge when English textbooks are full of page after page of exercises?

Motivation was a familiar theme a century ago: it was 'a well-known psychological fact that feelings of displeasure will facilitate fatigue while feelings of pleasure will check it'. (*Ibid.*:92) Interest is a crucial factor:

> This clue (to interest the child) must be sought in the child's nature. The child's instinct is to *do* — to exert his powers. Because to most adults pulleys are more interesting than permutations, it does not follow that they will be so for the child. Utility often acts powerfully in arousing the interest of the adult; utility alone is never the child's standard. When he appeals to it, asks what is the 'use' of this or that, he is sophisticated; he is not expressing his own views, but is appealing to the adult point of view. He never asks, 'What is the use of playing ball?' and this fact points to what is the child's standard of interest, namely *doing, successful* doing. (*Ibid.*:93–94)

Young continues:

> 'Liking mathematics' is practically synonymous with 'ability to do the work as presented'. The child does not object to 'puzzles', provided he can solve them: he likes them — witness the puzzle department seldom absent from children's papers or columns. 'The puzzle instinct' is not to be pooh-poohed, it is fundamental in the matter of the child's interest. The question of usefulness of content, whether as discipline or in practical applications, does not trouble him. The average child regards his teacher, his textbook, as sources from which legitimately emanate things for him to do, and if can do them, he is pleased and likes the doing. This includes purely intellectual doing as well as manual doing. (*Ibid.*:94–95)

We shall consider later how pupils can be abused by giving them things they can do but which are of no value: in contrast, we shall consider the old distinction between *problema* which indeed require something to be done, and *theorema* which require something to be proved.

Young also discusses another theme that features later in this book, the crucial importance of linking mathematics to other subjects:

> The laboratory method ... its insistent demand for a closer correlation of subjects, both of the mathematical subjects among themselves and of mathematics with physics. That arithmetic, algebra, geometry, trigonometry should be taught side by side by the same teacher to the same pupils, each helping and illuminating the other, and not tandem, as is the custom in America, has long been urged ... The laboratory method (also) proposes that the unified mathematics be brought into close relations with physics and with all the applications of mathematics ... The laboratory method makes the most radical proposition so far made anywhere, *viz.*, that mathematics and physics be organized into one coherent whole, the most extreme form of the proposition being that the reorganization be so thorough as to recognize in the secondary school no distinction between mathematics and its principal applications. (*Ibid.*:87, 99, 100)

Perhaps because I once taught both maths and science to the same primary pupils, I have always been strongly convinced of the benefits of linking these two subjects and the great loss to pupils when they are separated as they invariably are in British secondary schools.

Young does have weaknesses. When he discusses creating problems, it is always the teacher who creates new problems, never the pupil, though surely pupils in an actual *mathematical laboratory* must have often come up with questions which were also interesting problems. Yet the overwhelming impression of this book is how *progressive* it is! How child-centred! How reasonable! How rational! How much have we really progressed in the last century?!

1

Introduction: Motivation and Mathematics Teaching

> In general, I am not at all averse to the proposal that we should require
> less from the learner and that what he learns should be learnt better.
>
> Matthew Arnold (1910:213)

Mathematics is a notoriously hard subject that many adults are happy and
unembarrassed to admit they flunked at school and which many current
students avoid in making their examination choices, preferring 'easier'
subjects. So it is obvious, is it not, that *motivation* must be a key theme in
any discussion of school maths? Yes, and no. Yes, it is obvious but no it is
not a key theme, in fact it gets very little attention at all.

The Cockcroft report: *Mathematics Counts* (1982)

The famous Cockcroft Report is more than 30 years old but it is still worth
dipping into to remind ourselves of some of its highlights. It did not
explicitly discuss motivation. There is nothing in the index between 'modern
mathematics' and 'multi-base arithmetic', and nor is there any serious
discussion of the potentially motivating relationship between school science
and mathematics but it did have other pertinent points to make.

It did discuss attitudes and their origins (paragraph 198 *et seq*)
referring to,

> The fact that 'puzzle corners' of various kinds appear in so many papers
> and periodicals testifies to the fact that the appeal of relatively elemen-
> tary problems and puzzles is widespread; attempts to solve them can
> both provide enjoyment and also, in many cases, lead to increased math-
> ematical understanding. (*Ibid.*:7)

In contrast, Cockcroft reported the results of an especially commissioned
survey of adult numeracy. Many subjects refused to be interviewed, full
stop. Why?

> Both direct and indirect approaches were tried, the word 'mathematics'
> was replaced by 'arithmetic' or 'everyday use of numbers' but it was clear
> that the reason for people's refusal to be interviewed was simply that the
> subject was mathematics ... This apparently widespread perception
> amongst adults of mathematics as a daunting subject pervaded a great
> deal of the sample selection; half the people ... refused to take part.
> (*Ibid.*:16)

Cockcroft concluded that,

> the most striking feature of the study (was) the extent to which the need
> to undertake even an apparently simple and straightforward piece of
> mathematics could induce feelings of anxiety, helplessness, fear and even
> guilt in some interviewees. (*Ibid.*:20)

We might expect that this situation would have improved in the last 33
years, I hope it has, but any improvement is not reflected in more students
choosing to study mathematics at higher levels. We may also hope that
mathematics today is less painful but it is certainly found by many to be
difficult and worth avoiding.

Cockcroft drew attention to another major cause of low motivation in
two paragraphs which I make no apology for quoting at length:

> We wish now to draw attention to a further difficulty which arises for
> both examiners and teachers as a result of the fact that CSE examinations
> are not suitable for many of the pupils who now attempt them. Examiners

have a duty to set papers which cover as much of the syllabus as possible. Because they are aware that many low-attaining candidates will attempt the papers, they feel obliged to include within them a number of trivial questions on those topics in the syllabus which are conceptually difficult so that low-attaining candidates may find some questions which they are able to attempt. Teachers in their turn feel obliged to cover as much of the syllabus as possible so that their low-attaining pupils may be able to answer such questions, even though some of the topics which are included are conceptually too difficult for these pupils. This leads to teaching of a kind which, instead of developing understanding, concentrates on the drilling of routines in order to answer examination questions. We therefore have a 'vicious circle' which is difficult to break. (*Ibid.*:445)

Many teachers are aware of this problem but feel unable to do anything about it. Their dilemma was expressed vividly at one of the meetings which we held with groups of teachers. 'I know that I should not be teaching in this way and I would much prefer not to do so, but I know also that I have a responsibility to do all that I can to make it possible for my pupils to obtain a CSE grade.'

However, there can be little doubt that teaching of this kind can lead to disenchantment with mathematics on the part of many pupils ... (*Ibid.*:446)

Checking current textbooks and examination syllabuses suggests that this problem has in no way been solved in the last 30 years and more; if anything it has been exacerbated. The result is a massive demotivation of millions of pupils by the worst effects of an examination-dominated system in which even well-meaning teachers feel obliged to, let's not beat about the bush, dump their pupils in the ordure by deliberately training them for an exam rather than educating them in mathematics.

Cockroft's explanation for that disgraceful state of affairs was the obvious one, that syllabuses are first designed for the top-performing pupils and then watered down for the rest, 'by deleting a few topics and reducing the depth of treatment of others'. Cockcroft argued that, 'this is a wrong approach' and that development should be 'from the bottom upwards' (*Ibid.*:450).

Cockcroft concluded rather that,

> it should be a fundamental principle that no topic should be included unless it can be developed sufficiently for it to be applied in ways that the pupil can understand. (*Ibid.*:451)

adding by way of example,

> we see no value in teaching and examining, in isolation as a skill, the addition and multiplication of matrices to pupils whose knowledge of algebra and geometry is not sufficient for them to able to appreciate contexts within which matrices are of use. (*Ibid.*:451)

Cockcroft was also aware, so long ago, that problems of discipline which are often problems of motivation, affected the manner in which mathematics was taught:

> Especially as pupils become older, a great deal more time is often given to written work than to discussion and oral work. This situation very often arises from the fact that pupils who have become disenchanted with mathematics as a result of lack of success over the years can present problems of control in the classroom which make it difficult to continue oral work for any length of time. However, lack of discussion almost certainly leads to further failure and so the problem is compounded. (*Ibid.*:464)

Just so: it is worth noting that in many other countries, pupils are more motivated and better behaved and so extended classroom discussions are possible and are an important part of the lesson (see Ch. 12 of this book).

Cockcroft concluded that,

> Changes in the examination system and in the organisation of secondary schools ... in recent years have influenced the teaching of mathematics in ways which have been neither intended nor sufficiently realised. At the present time up to 80% of pupils in secondary schools are following courses leading to examinations whose syllabuses are comparable in

extent and conceptual difficulty with those which twenty years ago were followed by only about 25 per cent of pupils. Because ... it is the content of O-level syllabuses which exerts the greatest influence, it is pupils whose attainment is average or below average who have been most greatly disadvantaged. (*Ibid.*:442)

This is as valid today, and as shameful in its implications, as it was then. What have we learnt?

The Smith report: *Making Mathematics Count* (2004)

Professor Adrian Smith's (2004) report of his inquiry into post-14 mathematics education, whose title is a play on Cockcroft's, identified as a key issue,

the failure of the current curriculum, assessment and qualifications framework in England, Wales and Northern Ireland to meet the needs of many learners and to satisfy the requirements and expectations of employers and higher education institutions. (*Ibid.*: paragraph 0.13)

The summary of Ch. 1: 'The Importance of Mathematics' noted that there is 'a long-term decline in the number of young people continuing to study mathematics post-16' (Scotland excepted) and listed these four factors:

- the perceived poor quality of the teaching and learning experience;
- the perceived relative difficulty of the subject;
- the failure of the curriculum to excite interest and provide appropriate motivations;
- the lack of awareness of mathematical skills for future career options and advancement.

The report also emphasised other problems: 'There is widespread agreement that the Key Stage 4 curriculum is overcrowded ...' (*Ibid.*:0.28) making it impossible to teach all the topics well or to get most pupils up to the level they could and should reach if the curriculum were smaller.

Professor Smith also believes that, 'we do not sufficiently stretch and motivate the top 10%' (*Ibid.*:0.29). There's a novel thought, but why does he only refer to the top 10%? Is he suggesting that the rest, no less than

90%, are sufficiently stretched and sufficiently motivated? I don't think so. So why this distinction?

Are the top 10% more worthy of special attention than the bottom 10%? Yes, no doubt, if you are an industrialist but not if you are a teacher. If children are forced by law to attend schools where they are then forced by law to attend maths lessons where the curriculum too often fails to excite their interest and does not provide appropriate motivations, and where the content of the course may seem boring or irrelevant, or insufficiently stimulating or challenging (*Ibid.*:1.11) whose fault is that, and who is responsible for doing something about it? One of the arguments of this book, implicit more than explicit, will be that all children can and should be motivated to succeed at mathematics, within their capacities, which I would confidently assert are much higher than usually admitted or recognised. But they will never do so under the present dispensation.

Mathematics is, of course, seen by many pupils and their parents as difficult relative to other subjects:

> In the UK there is a widespread view, among both parents and students, that the subject itself is 'difficult' and 'boring' and presents disproportionate challenges in the school and college setting. (*Ibid.*:1.17)

Since it is also considered by respondents to Adrian Smith to require much effort, equivalent to double science, Smith recommended that mathematics be counted as a double award.

Smith also notes that the Foundation Tier of GCSE Mathematics awards grades D, E, F and G. Since C is regarded by almost everyone as a 'pass', pupils who are put in for the Foundation Tier exam (30% of the cohort!) are destined to fail even before they sit the examination (*Ibid.*:0.26, 3.17). As Professor Smith put it to a Commons Select Committee, employers still assume,

> That a C is the threshold gold standard for 'success' at GCSE, so if not getting a C is a failure and you arrange a tiered system so that something like 30 per cent of the age cohort are entered for a tier where they cannot acquire that which is perceived to be success you have something really odd and *quite distressing*.

> (Smith 2004b, viii, italics added for emphasis)

Only in a very punitive and judgmental system and society would 30% of pupils in any subject be classified as failures long before they take the relevant exam. In a rational society, the blame for their 'failure' would be laid where it belongs — at the feet of their teachers, textbook authors, professional examiners and syllabus and curriculum designers, and of course successive governments.

We might quote Cockcroft again here:

> We cannot believe that it can in any way be educationally desirable that a pupil of average ability should, for the purposes of obtaining a school-leaving certificate, be required to attempt an examination paper on which he is able to obtain only about one-third of the possible marks. (Cockcroft 1982:444)

Indeed, no. Motivation is maximised when pupils more or less master the material: it is seriously reduced if pupils expect to end a course with only a so-so grasp of some, only, of the topics on the syllabus. (And examination success is maximised when pupils take an examination when they are ready for it, not when the teacher decides.)

We may also note that adults are usually only failures at playing chess or the violin if they initially *volunteered* to *attempt to succeed*. In real life, children and adults are selected upwards according to their success and so failure is minimised. In school, children are selected downwards by their failure to reach standards which their teachers know all along many of them would fail. Thus failure is maximised and pressure is put on children which is used to control them and to judge them (1981a:37).

Finally, Smith concludes that the problems of teaching mathematics are exacerbated by a specific and daunting problem: 'We think it likely that there is a current shortfall of around 3,400 specialist mathematics teachers in maintained secondary schools in England' (Smith 2004:0.16), and Smith notes that according to a recent survey, 'over 30% of those currently teaching mathematics do not have a post A-level qualification in mathematics' (0.16), while 25% of teachers qualified to teach mathematics are doing something else (0.17).

No wonder that motivation suffers and the image of mathematics with it.

A Nuffield perspective

A 2004 report from the Nuffield Foundation also criticised the state of affairs:

> **Impoverished experience of mathematics.** The frequency and style of assessment, currently practised, encourage short-term objectives and teaching methods which impoverish the experience of mathematics. Portfolios of work reflecting problem-solving competency would be one way forward.

An 'impoverished experience of mathematics' must reduce motivation whereas problem solving does indeed raise motivation, as I shall argue below. The authors next argue that:

> **Motivation** in terms of relevance and utility too often overshadows the equally motivating force of interest (at all levels) in mathematical problem-solving. And, indeed, what is seen by students to be 'relevant' is not as easy to determine as is often assumed.

> (Nuffield Foundation 2004;6, 20)

Once again, I am happy to agree, and add that the motivating force of puzzles is very well known, although most puzzles have no practical use or benefit whatsoever — apart from forcing the solver to think which is just what mathematics teachers should want them to do!

The authors add that, 'At present, backed by Ofsted inspections, the curriculum is circumscribed and limiting' and that while 'Modules seem to be [an] obvious organisational framework within which flexibility of pathways can be managed', they are open to 'serious objections. A richer mathematical experience would find room for trial and error, reflection, the struggle to understand, tentative conclusions — the kind of learning which short-term modules ... obviate.' (*Ibid.*:6)

Finally, they repeat the by now familiar refrains that,

> There is growing evidence that mathematics, as it is currently formulated, is becoming increasingly unappealing to a substantial number of students (*Ibid.*:20)

and that,

> Mathematics is often seen as a difficult subject to study when compared to others, and as it becomes more abstract is perceived to have little relevance to students (*Ibid.*:21)

and they conclude that,

> The current assessment burden is too great for students, and is also detrimental to their experience of mathematics. (*Ibid.*:22)

Mathematics is invisible

> Mathematics enjoys the advantages, and suffers the penalties, of being a secret doctrine.
>
> (M.H.A. Newman in Cartwright 1955:37)

It is popularly said (among maths educators, that is), that 'mathematics is not a spectator sport', meaning that pupils need to get stuck in and actually do mathematics in order to understand it. Indeed. However, this proverbial saying has another very different meaning: mathematics is not a spectator sport *because it is invisible.*

This is very demotivating and yet another reason for the difficulty of mathematics. Mathematics *as a practice* is invisible when compared to activities such as football or other sports, or acting or even business, let alone the visual arts. Television programs, in between endless so-called 'reality' shows, do sometimes show painters painting, climbers climbing and actors acting. Who sees a mathematician doing maths? If you could see Andrew Wiles doing mathematics, what would you see? Some geeky figure writing across a blackboard? No, that is a cartoon caricature promoted by Hollywood films.

It takes a connoisseur to appreciate the finer points of the action on the pitch but the ordinary knowledgeable spectator can appreciate much of the action and mere tyros (like me) can marvel as Rooney chips the ball over the head of the advancing goalkeeper or Ronaldinho slips the ball between a defenders legs and picks it up on the other side.

No wonder that kids can appreciate football so much, or tennis, or rock climbing if that is their interest, yet often find the simplest maths baffling.

One major reason why mathematics is difficult is that it is more or less invisible. We shall return to this theme when we discuss the appreciation of mathematics in Ch. 4.

Motivation and the education of mathematics teachers

The ERIC (Educational Resources Information Centre) database, which is available free on the internet, does not list 'motivation' among the 54 descriptors in its mathematics thesaurus though it includes 'mathematics anxiety', suggesting that it is more concerned about the negative effects that mathematics teaching can have on pupils than with the positive effects that might make mathematics more acceptable in the first place. Motivation is not, it seems, uppermost in mathematics educators' minds. Even more surprisingly, motivation does not feature strongly in books about the teaching of mathematics.

There are exceptions: here are five. Mason's *Fundamental Constructs in Mathematics Education* devotes a whole chapter to motivation in which he discusses the views of Mary Boole, Launcelot Hogben and Richard Skemp plus Nitsa Movshovits-Hadar on surprise and Alain Bouvier on the role of disturbances or ruptures, among many other themes (Mason 2004:Ch. 4), while *Developing Thinking in Geometry* by Susan Johnstone-Wilder and John Mason (2005) has an excellent section on 'harnessing emotion'.

Another exception is *Mathematics and Motivation* (1995) edited by Martha Carr, which approaches the subject from the opposite direction to mine, from academic psychological research rather than classroom experience. *Rethinking the Mathematical Curriculum* by Hoyles, Morgan and Woodhouse (1999) is unusual in discussing 'aesthetic appreciation, cultural knowledge and creativity':

> Teachers in primary schools know that young children find pleasure and emotional engagement in the aesthetics outcomes of their observations and activities ... Reasoning develops along side aesthetic appreciation and emotional response ... Later still ... aesthetic appreciation develops to become rather more intellectual than sensual, focusing, for example, on elegance of abstract structures or economy of forms of reasoning.

Similar transitions from the sensual to the intellectual also happen in the arts subjects ... As the mathematics curriculum becomes more analytical, teachers need to remember the importance of retaining an aesthetic dimension ...

(Hoyles et al. 1999:20–21)

Finally, Goulding in *Learning to Teach Mathematics in the Secondary School* (2003) gives a long list of her own aims, and therefore motivation, according to the demands of different situations. Here are some of her points which especially appeal to me:

- I want you to puzzle this out for yourselves.
- You will probably need this fact/skill/way of thinking at this point.
- Because this is a very useful thing to know.
- This is a really powerful idea and it took human beings a long time to develop it.
- I want to share my own fascination with you.
- This is very attractive and pleasing.
- I do not know how to solve this myself.
- You need to prove that your idea will always work.
- You will all learn more if you work together.

She concludes:

When you can share your intentions with pupils in this way, it is usually much easier to teach. The pupils have a clearer idea of the agenda, and will know what to expect and what not to expect.

In other words, the teacher is deliberately putting the pupils 'in the picture' (Goulding 2003:131). Such analyses, however, are exceptional.

2

Demotivating and Demotivated Pupils

What's the point?

Many years ago I was introducing a new topic from one of the *School Mathematics Project* textbooks when a boy piped up: 'What's the point of this, sir?'

It was not the first time I had heard the question. On this occasion, however, I took it to the other mathematics staff at midmorning break and asked them what they did when faced with such demanding and inquisitive students. The Head of Department claimed he had never heard the question which I found difficult to believe. The deputy head of department who had the unusual distinction for a maths teacher of producing and directing the yearly school play, recognised the question and replied that he told such pupils to 'get stuffed!', though not, of course, in precisely those words; the third member also recognised the question but admitted he had no good answer to it.

Fast forward 30 years. The *Education Guardian* for 22 October 2002 had a lead article, 'Sum problems. The search for a new formula to solve the maths crisis'. John Crace reported that he is repeatedly asked by his maths pupils, 'What's the point?' Well, there's a novelty!

Fast forward again and in a *Guardian* supplement, *Magic in Maths* (27-11-07) Victoria Neumark referred to 'the dreaded question 'What's the point of this?"

How terrible that such a natural and potentially innocent question should be dreaded by any teacher — and how absurd that the question is not automatically answered, convincingly and simply, by any textbook! Unfortunately, it often cannot be answered *honestly* because an honest answer does not exist or would be too embarrassing to produce.

One factor which undermines pupils' enthusiasm and willingness to work hard, or indeed to work at all, is lack of meaning: they don't 'see the point' of what they are doing. Typical textbooks don't help.

Ideally, pupils need to appreciate that what they are learning can be used for some worthwhile purpose, that they can do something with it — solve a problem — which they themselves find puzzling and therefore interesting and engaging. They need to be active learners, not just by learning actively but by learning ideas which have an active use rather than just lying passively in their memory banks to be drawn out at exam time, and then forgotten.

The honest answer to the original question will often be,

> There's no point for you, Peter, because you are best at drama and English and you are going to specialise in those subjects later on. We have to do this topic for the benefit of John and Mary who are going to study science at A-level.

However, that is not an answer that Peter ought to accept — that he is studying a topic not for his own benefit but for someone else's — and it is certainly not an answer that most teachers would dare to give.

Closely related to this theme is the following point from evidence submitted to the Cockcroft Committee:

> Mathematics lessons in secondary schools are often not about anything. You collect like terms, or learn the laws of indices, with no perception of why anyone needs to do such things. There is excessive preoccupation with a sequence of skills and quite inadequate opportunity to see the skills emerging from the solution of problems. As a consequence of this approach, school mathematics contains very little incidental information. A French lesson might well contain incidental information about France — and so on across the curriculum; but in mathematics the incidental information which one might expect (current exchange and interest

rates; general knowledge on climate, communications and geography; the rules and scoring systems of games; social statistics) is rarely there, because most teachers in no way see this as part of their responsibility when teaching mathematics.

(Cockcroft:462)

Cockcroft then points out that this argument is by no means limited only to courses for low-attaining pupils.

A nasty experience

I recently had a distressing experience: I was introduced to an American textbook, *Algebra 1*, third edition 2006, by Paul Foerster, published by Prentice Hall who describe it on the cover in a golden-style medallion as a 'Classics Edition'. The 'Foreword to the Student' is nothing if not frank:

> Some of the things you will learn may not seem to have an immediate practical use. Learn them anyway, and learn them well! They are all part of a big picture which becomes clearer only after you have unveiled its various parts.

I was amazed to read this confession of a stance whose defects should have been obvious to the author in 1961 when he began to teach at Alamo Heights High School in San Antonio, Texas.

There are two giant defects in this approach:

- If pupils cannot see the point of what they are learning today, then their motivation will suffer and so their learning will suffer too. It is hard enough for adults to perform well when their supposed goal is located mysteriously in a distant and obscure future, and far harder for children.
- Claims that what you are learning may seem pointless now but *all will be revealed in good time* invariably turn out to be spurious.

Sure enough, turning to the end of this 738-page text, the Big Picture still isn't there — so that's one year gone by with minimal illumination. British textbooks are not so brutally honest, but the spurious claim is there, albeit implicit not explicit, and it also turns out to be false. Do British textbooks give their pupils the Big Picture to start with, or do they also present bits

The Big Picture

A man wishing to learn to drive applied to a professor of engineering for instruction. On Monday the professor explained the theory of the heat engine. On Tuesday he showed how the internal combustion engine exemplified the basic principles he had expounded on Monday. On Wednesday he examined the different parts of the engine, explaining the function of each part separately. On Thursday he described with the aid of an exploded diagram of an internal combustion engine exactly how the various parts worked together to provide the motive power for the car. On Friday he was called away to an important meeting and when the weekend came the unfortunate learner was unable even to drive his car out of the garage.

(1977:15)

of this and bits of that, here a piece, there a piece, but never bringing the pieces together to complete the jigsaw? The latter, unfortunately.

A major reason why the question, 'What's the point, sir?' is not anticipated and pupils are not told why they are being taught a topic, is that *for them* there is no reason, apart of course from finding the topic on an exam paper.

Mathematics textbooks tend to focus far too much on explanations followed by exercises and more exercises, tempting teachers to follow the same route. Textbooks also have many other striking defects of which we will mention two here: the lack of reference to science, and lack of reference to proof.

Science in mathematics textbooks: science as a motivation

When the GCSE was yet still young, the General Criteria required,

> that all syllabuses should be designed to help candidates understand the relationship of the subject to other areas of study and its relevance to the candidates' own lives.

> (*Secondary Examinations Council* 1986)

That quotation is taken from *Making Science Education Relevant* (Newton 1988) a book which, ironically, does not list mathematics in its index.

One Big Picture for arithmetic and algebra, as all scientists and all mathematicians know perfectly well — and the biggest Big Picture for all but the pure mathematicians among us — is that maths provides many of the concepts, methods and language of the hard sciences without which you cannot do physics or chemistry, engineering or architecture, not forgetting astronomy, biology, financial modeling and the uses of statistics in the social sciences.

The profound historical and practical links between mathematics and the hard sciences are an obvious source of motivation which is hardly exploited at all in teaching school mathematics.

I will add that I benefited greatly by starting teaching science and mathematics in a progressive primary school to 9 to 11 year olds: an open invitation to use simple science equipment in maths classes as well as the usual primary apparatus augmented by assorted fascinating objects such as large and very heavy gear wheels from a parent who ran a building business, free samples of ceramic tiles of assorted interesting shapes from a local shop, and so on.

Look at most mathematics textbooks — or at examination syllabuses — and it is apparent that there is little or no contact between maths and other subjects, although the quantity of mathematics used by other subjects is increasing all the time.

Some small steps are currently being taken to remedy this situation, notably through the STEM (science, technology, engineering and mathematics) initiative but it will be a long time before such concepts and attitudes are taken for granted in the ordinary classroom.

Absence of proof for most pupils

Mathematics is a subject — the subject — in which argument can be very precise and convincing. Such arguments, when the subject matter is of some interest and not merely routine, are called proofs, and you might suppose that the idea of proof would be central to the pupil's study of mathematics.

Not so. Pupils are often encouraged to discuss and argue but the idea of *proof* is apparently so difficult, so advanced, so hard to handle, that it must be left for most pupils until almost the very last moment, if they ever reach it at all.

I shall discuss proof in primary and secondary schools at length in Ch. 6. In the meantime, here are the various views of the National Curriculum on the proper location of proof for secondary pupils. They do not suggest confidence in pupils' abilities, or high expectations.

Proof in the 2008 National Curriculum

During Key Stage 4 (foundation)

'They begin to understand and follow a short proof.'

During Key Stage 4 (higher)

'They learn the importance of precision and rigour in mathematics.'

'They use short chains of deductive reasoning, develop their own proofs, and begin to understand the importance of proof in mathematics.'

STANDARD 7: REASONING AND PROOF

'Mathematics instructional programs should focus on learning to reason and construct proofs as part of understanding mathematics so that all students:

- recognise reasoning and proof as essential and powerful parts of mathematics;
- make and investigate mathematical conjectures;
- develop and evaluate mathematical arguments and proofs;
- select and use various types of reasoning and methods of proof as appropriate.'

Anxiety

Syllabuses and textbooks are, however, very well designed to induce anxiety in many pupils and so damage their motivation and cause them cognitive problems too.

We have already noted that 'mathematics anxiety' is a key term in the ERIC thesaurus, though 'mathematics motivation' is not. We can add that *no other subject* has this privilege. Do physics, French and history sometimes make their pupils anxious? No doubt. French made me very anxious because I had a hopeless memory for French irregular verbs and French vocabulary in general which might as well have been more-or-less random syllables strung together. But is 'French anxiety' a recognised phenomenon? No.

Mathematics is unique in its capacity to arouse extreme anxiety, as Laurie Buxton documented in his book *Do you Panic about Maths?* based on his own research with adult subjects including the successful headmistress of a large school who had a degree in English but started perspiring and palpitating when faced with the simplest maths problem.

Given this unique potential, anyone might assume that all teachers would be extremely wary of putting pupils into situations which might make them anxious, first, because it is morally suspect to expose pupils for whom we are *in loco parentis* to distress; second, because anxiety strongly interferes cognitively with learning and so undermines all our efforts to teach the anxious child.

According to Ashcraft and Ridley (2005), mathematics anxiety leads victims to avoid mathematically laden situations, to play down the usefulness of mathematics, and to value their mathematics teachers less. Severe anxiety also interferes with working memory (Campbell 2005:315–327).

Teachers of anxious pupils cannot but observe their tendency to jump at an answer — any answer — rather than pausing to think calmly and *confidently* about the problem, another consequence that Ashcraft and Ridley observed in their experiments: highly anxious subjects often answered as quickly as calm subjects but made far more errors.

So, yes, we might assume that teachers would avoid anxiety-laden situations, but, no, they occur very frequently.

Brian Butterworth, in discussing *developmental dyscalculia* (DD), gives this transcript to illustrate the emotional effects on three 9-year-old children who find that they are 'struggling with mathematics tasks that (their) peers find very easy':

Child 5:   It makes me feel left out, sometimes.
Child 2:   Yeah.
Child 5:   When I like — when I don't know something, I wish I was like a clever person and I blame it on myself ...
Child 4:   I would cry and wish I was at home with my mum and it would be — I won't have to do any maths.

In another experimental group, Child 4 is smarter but recognises the distress of a girl who is failing:

> Child 4: Yeah, and then she goes and hides in the corner — nobody knows where she is and she's crying there.

Anxiety, says Butterworth, does not cause DD but DD certainly causes anxiety, as does failure in general in mathematics, for whatever reason, for many pupils (Campbell 2005:315–372).

Moreover, failure at mathematics can appear very suddenly as all teachers and parents know: it is only necessary to fall behind on one topic, lose confidence, maybe find that the next topic doesn't make sense because of the new gap in their understanding, and the pupil becomes anxious and potentially enters a downward spiral.

'Anxiety, insight and appreciation' (1994c) presented the following anecdote from a seminar on teacher attitudes:

> (the leader) described some research by a student of his, Richard, who found one of his pupils becoming very anxious, panicking. He took her aside and talked to her and her anxiety subsided. The (seminar leader) went on to describe how, over a period of time, the girl learnt to recognise when she was starting to panic, and to warn Richard in advance. When I objected that Richard was effectively including this panic and should have been confronted with this fact, no one else agreed, and there were actual objections.

> The first objection was that the girl was lucky to have such a sympathetic teacher — which is like arguing that if the driver who knocks you down is a doctor with a beautiful bedside manner, you're a lucky victim, which is true but you are still a victim and nothing justifies the doctor hitting you in the first place. The second objection was that most pupils panic anyway, so why make such a fuss? Because severe anxiety, panic even, is a very distressing phenomenon, widespread within mathematics education … linked to the reputation of mathematics as a very difficult subject …

Compare the reactions illustrated in that anecdote with this quotation from *Ethical Standards of Psychologists*, published by the American Psychological

Association as long ago as 1953:

> Principle 4.31–1. Only when a problem is significant and can be investi-gated in no other way is the psychologist justified in exposing research subjects to emotional stress. He (sic) must seriously consider the possi-bility of harmful side-effects and should be prepared to remove them as soon as permitted by the design of the experiment. When the danger of serious side-effects exists, research should be conducted only when the subjects or their responsible agents are fully informed of this possibility and volunteer nevertheless.

Why is it acceptable for the education system, via its examiners and text-book authors and teachers to allow distressing levels of anxiety to develop in so many pupils, when psychologists are not allowed to do the same? Is mathematics so important, and can it be only taught in such a way that it 'exposes (so many pupils) to emotional stress'? How can we maximise pupils' motivation if so many pupils are anxious?

Syllabuses crammed for the sake of stronger pupils

One feature of the usual syllabuses that poses problems for teachers and pupils is that they are overcrowded. We have noted Adrian Smith's conclu-sion on Key Stage 4. *Making Mathematics Count* also suggests (Smith:0.28) that there is far too much statistics and data handling and that some it should be shifted onto other subjects such as geography and biology, leav-ing pupils more time to master the core topics and themes.

There is a reason for all this mass of material, of course: it is there for the benefit of the stronger pupils (who may possibly need it later for their more advanced studies) and industrialists and university dons.

No doubt the textbook authors are not entirely to blame. They are providing what the publishers demand, the publishers publish what teach-ers demand in order to get pupils through public examinations, which are in turn hamstrung by syllabuses and a curriculum decided, in the final analysis, by the government. The result, ironically, given the government's enthusiasm for improving the quality of mathematics, is more pressure for all, more anxiety for many, and the destruction of motivation.

Time pressure and lack of time

It is greatly to the convenience of teachers and administrators that each pupil should work through the subject matter of a syllabus at the same rate and eventually take the same examination on the same date.

Traditionally, time pressure has been laid on, and the convenience of adults served, by following a syllabus which most children have no chance at all of mastering because it contains, as we have noted, far too much material. Most pupils will never have the chance to become familiar with its contents, to feel at home with it and in control of it, rather than being pushed around by it.

The overloading of the syllabus has serious effects on all pupils but especially on the weaker who will be stressed and strained by trying to keep up, if they have not already lost confidence and fallen behind, in which case the blame for their failure is traditionally laid squarely on their narrow and frequently blameless shoulders.

The inevitable result of this pressure of time/content is that very few children ever enjoy the feeling of mastering anything. The pleasure and

Time pressure

'Schools traditionally do not allow or expect pupils to be creative in mathematics, in contrast to other subjects like English or art, and even when children are given the opportunity, the usual constraints of schooling tend to tie the pupil hand and foot. In particular, the fact that insight into a problem takes time, perhaps a lot of time, during which the pupil may produce nothing which can be marked or put in his or her folder, clashes violently with the division of the school day into quite short chunks of time, 90 mins at the very most, and the usual emphasis on success. How is it possible, within the usual system, to give a child credit for an insightful attempt, spread perhaps over several lessons, to solve a difficult problem which is ultimately unsuccessful and does not produce a final answer?'

'If it isn't possible now, then we must make it possible, because professional mathematicians are often in just such situations, as indeed are professional artists or scientists. Children cannot model themselves on the professionals if they are not, sometimes, given the chance to fail gloriously.'

(1983:8)

satisfaction of feeling that you are in control of the material is completely unknown to many children, though it is one of the greatest motivators of academic work.

Pupils need time to explore, time to experiment, even more time to produce convincing arguments and proofs and more time still to develop an *intuitive* understanding of each topic.

They also need time to occasionally fail.

So what is the teacher to do when faced with both pupils' (implicit) demands for more time *and* the pressures of examination syllabuses and the textbooks which service the examination industry? We can sketch a continuum along which the teacher tends to move. In three stages it goes like this:

(A) Loads of time for experiment and exploration, and loads of time for discussion and argument, loads of time to solve problems and loads of time to write about what you've been doing.

(B) A little time for experiment and exploration, a short discussion led by the teacher, an exercise from the textbook with genuine problems only appearing at the end so that only the strongest pupils will reach them. No time to write up what the pupils have been doing, because actually they haven't been doing anything to write about.

(C) No time for experiment and exploration, the teacher introduces the topic from the textbook, explains it using the textbook, and then sets an exercise, from the textbook. The textbook author 'helps' pupils who have not grasped the explanation by providing rules which pupils can use to answer the exercise. No problems to solve, nothing to write home about, indeed nothing to write about full stop. And, of course, minimal understanding, if any.

This Rake's Progress can be seen from two perspectives: first, as the progress typically seen from early primary and then all the way up to GCSE classes: as the years go by the syllabus is more and more over-loaded and pupils have less and less time to grasp anything deeply.

Secondly, as the progress of one teacher, from initial enthusiasm and determination (just out of training college) to give pupils real understanding and insight, to the realisation that if you start the year with that approach, you will be way behind within weeks, and hence to the conclusion that there really is no time for pupils to develop their understanding naturally, and that force-feeding them pap is the only way to get a proportion of them through their public exams.

Vicious circles of decreasing expectation (and motivation)

Many pupils face another very demotivating experience. On transfer to secondary school or later, perhaps more than once, they are dropped down into a lower stream or mathematics set on the grounds that they are not good enough to 'keep up' with their smarter classmates.

In 'theory' as supporters of classification and categorisation will assure us, this is a good thing because it guarantees that they will be studying with other pupils of their own ability from appropriate textbooks aided by sympathetic teachers.

In practice, the lower sets know perfectly well why they are there — it's because they are 'thick'. They usually do not get the best teachers — (in one of the many schools with a disastrous shortage of qualified and experienced maths teachers, who will the less qualified teach: the top GCSE set?) and the expectation is that they will do badly whoever teaches them, so frankly, what's the point?

All too often the situation is even worse. The process goes like this:

1   John Smith is doing badly in Set N, for one of two plausible reasons:

A   The work is too hard.
B   The work is boring or simply badly taught.

2   The teacher almost certainly selects explanation A, if only in self-defence. Everyone knows that many pupils are 'thick' and no one wants to admit that their lessons are tedious or incompetent.

3   So John Smith is put down a set into Set N+1 where the work is 'easier' — but he still does badly, for one of two plausible reasons:

A   The work is still too hard because John really is bad at maths.
B   The work, being 'easier', is even less challenging and more boring than before.

4   The teacher, possibly after consultation with colleagues, once again selects explanation A. Everyone in the department knows by now that John Smith is bad at maths, so 'A' is the 'obviously' correct conclusion.

5   So John is 'correctly' dropped down into Set N+2.

6   Where he is very bored indeed and does very badly indeed.

This vicious spiral downwards to the bottom matches a spiral of low and decreasing expectations followed by many pupils. Some of them will genuinely find mathematics very hard — but will not necessarily do better as a result of being dropped, because of the negative effects of being in Set Z.

Other pupils may simply be slower and need more time to master the concepts, but as we have seen, time is just what many pupils don't get and cannot get under the present dispensation.

We can illustrate this possibility by a comparison with children who are learning chess or go. One of the strongest British go players of several decades ago had a learning curve that started to rise very slowly so that he took a long time to reach his potential which turned out, however, to be very high. He eventually became one of the strongest British players. In contrast, several other players had learning curves which started more steeply and then levelled off, reaching in the end the same grade or even a lower grade. Needless to remark, if he had been taught in school he would have quickly been labelled 'C-stream', and might well have given up the game in disgust. Also needless to say, for most practical purposes, for example in most jobs, it is the final stage that is important, and not the shape of the learning curve.

Chess and go players will have, of course, different cognitive styles. Some will be more visual, others less so, some will think faster than others and so on. What happens when they are given as long as they need to learn chess with no pressure to succeed quickly? The answer can be, or should be, very surprising to mathematics teachers. (See box overleaf.)

In the final figure, A and B don't understand the subject perfectly but they do understand it well enough to answer exercises pretty well and to recognise the ideas when the topic is repeated next year in a classic 'spiral of learning' curriculum, and have 'obviously' done far better than C and D who are 'clearly' too weak to ever master the topic, and out of kindness to

Taking the pressure off

Here is a likely outcome if four beginners learn a game together:

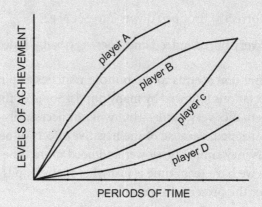

Different routes to success

The bottom scale shows time passing, and the side scale the rising level of achievement (...) Player A learns quickly and is soon soaring away. Player B learns quickly but then slows down perhaps because she has a poor memory and doesn't calculate that quickly. Player C learns more slowly, but eventually catches up with B. Perhaps he has a good memory which helps him to analyse positions (once understood) quickly and accurately. Player D starts and continues slowly.

The first point of the graph is that after a considerable time, players B and C have reached the same level of performance, albeit by very different routes!

This is a realistic picture of what happens when some children learn abstract games. Apparently very different children can reach the same level of final competence. Contrary to first impressions, speed of learning, quickness of thinking and good memory do not correlate 100% and many pupils with different talents can reach high levels of skill, given time.

This is an important point for maths teachers because most pupils are not aspiring to be pure mathematicians but only need maths for practical purposes and if they can eventually reach the required level of skill then they will have achieved all they desire. So if C and D are allowed more time, then even D may reach a worthwhile level of skill and C might indeed match B.

(Continued)

(*Continued*)

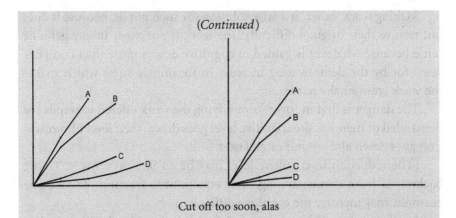

Cut off too soon, alas

In the left-hand figure, however, although B and C seem to be doing moderately well, the possibility that C might ever match B already seems unlikely: it has been 'cut off', literally.

For many pupils, however, even this time-extension is 'not allowed'. Instead, the whole class is cut off at the point when a new syllabus topic is introduced, many pupils are left with a very limited understanding that they have never been given time and experience to develop to anything near their full potential, and so we get the graph on the right-hand side.

(Edited from 2008c; based on 1979a:19–20)

their blancmange brains should be given easier work to 'master' and maybe moved down a set or even a stream until they fall, in a sad inversion of the famous Peter Principle, to their own 'level of incompetence' (2008c; 1979c:19–20).

Yet other pupils could simply be bored because *the work is boring*: they aren't challenged, they aren't expected to think for themselves, and they find the task of reading the teacher's mind beyond them and of no interest anyway.*

---

* Many extrinsically highly motivated pupils will try very hard to 'read the teachers mind' and they often come up with ingenious pseudo-concepts in place of the correct concepts they should be learning: such pseudo-learning may enable them to survive for the time being — until it eventually becomes apparent that their understanding is built on sand and needs to be redeveloped from the foundations up.

Making work 'easier' is a false solution for such pupils, because it does not resolve their original difficulty: the work is *not* easier in any genuine sense because whatever is gained in cognitive ease is more than compensated for by the demotivating increase in meaninglessness which makes the work emotionally harder.

The danger is that in (over-)simplifying the work offered to pupils (or demanded of them) as the cognitive level goes down, their level of motivation goes down also — and often faster.

(The reduction in cognitive level may be an illusion anyway. 'Going backwards' from elementary algebraic equations to fractions, to take one example, may increase the cognitive difficulty.)

I will add that when a class reaches this bottom level with very low cognitive demands in the absence of motivation, the solution to the problem is to make the work harder *but,* a crucial *but,* not by teaching from the front of the class which is almost certain to fail with such low achieving and demotivated pupils, but by giving the pupils challenging problems to work on themselves, and the time to think about them — just what the same pupils should always have been getting if they were to be maximally motivated!

'Maths is easy'

One response to the real difficulty of mathematics by teachers who are not prepared to tell pupils the truth is this popular lie, told with the best of intentions but with disastrous consequences: mathematics is, er, not very difficult. Even some maths educators have supported this claim:

> If we know that ineffective teaching of mathematics is not due to the difficulty of the subject matter ...

> We still have substantial illiteracy, widespread incompetence and incomprehension in all areas of a curriculum which, God knows, is not particularly demanding. The best students we have are said by their teachers to be in many cases unskilful in performance and superficial in understanding.

> (Wheeler 1970)

This is a complete misjudgement of how mathematics is subjectively perceived and experienced by almost all pupils. Mathematics is hard, and if you pretend that it is easy then pupils, not being fools or silly, won't be taken in by claims that their own experience tells them are false.

If you claim that mathematics is easy then the achievement of the strongest pupils is undermined, while those pupils who find it hard are put down and insulted by the implication that their difficulties are due to their own stupidity rather than the genuine difficulty of the subject and that whatever they do achieve counts for little anyway because it was, after all, 'easy'. Motivation and reward are *minimised*.

Meanwhile, any temporary increase in motivation from pupils who believe — for the time being — the claim that maths is easy, will soon be lost and their long-term motivation will suffer.

In contrast, by admitting the truth, but also claiming (correctly) that all pupils can do a lot of mathematics and be successful at their own level, pupils are motivated to try harder and all their achievements are *maximised*.

Distributing blame

What happens when the child's performance is judged to be not up to scratch? How is the blame distributed?

The child probably gets a report once a term and certainly once a year. This report will blame only one factor, the child ...

Look at a group of children, however, and the picture changes. John Smith is no longer the only failure, he is not even one of a small minority but of a large majority who perform poorly if not downright badly in the terms *set by the educational system itself*. Clearly some of the blame does lie with individual children, their parents and their teachers. But looking at the result of all children together, the system must be at fault itself ...

But in that case, what is the moral justification for blaming children for all of their own failure, when the failure of the system is not admitted? Isn't that completely dishonest? And is not honesty one of the virtues that the system is supposed to prize? Is not the system supposed to promote the very virtue which it treats with contempt?

(1980:22–23)

There is a moral angle here also. All education can be interpreted as a moral activity, but mathematics more than most, because of the massive levels of failure that it creates. It cannot be moral, nor can it be professional, to demoralise pupils.

Of course, having explained to pupils that mathematics is difficult, it is up to the teacher to ensure that they are successful most of the time. (And that they appreciate that professional mathematicians do spend much of their time more-or-less failing, and that they accept this.)

Sets that are too simple

There is another side to the 'maths is easy' claim. *Motivating topics cannot be too simple.* To put that another way, it takes sophistication to appreciate that — for example — set theory is actually very subtle and interesting. In the meantime:

> General set theory is pretty trivial stuff really, but, if you want to be a mathematician, you need some, so here it is; read it, absorb it and forget it.
>
> (Halmos 1974)

That was Paul Halmos, a professional mathematician who has also won awards for exposition, introducing undergraduate students to his own book on basic set theory. School students cannot be expected to show the same level of *appreciation*.

School textbook authors do not take Halmos's tip and do not point out to pupils that 'sets' as they appear in their textbooks are really jolly simple, have very few uses in the rest of secondary maths (or science) and (at most) have to be learnt before you move on to more exciting and challenging topics.

'Have to be learnt?' Yes, but only because they are in syllabuses and hence in textbooks and examinations. In a more realistic world, any slight ideas of sets needed at primary and secondary level would be introduced *surreptitiously* and no fuss would be made about them.

The notation of sets is almost useless at school level, certainly below 16–19, and the school topic of 'sets' makes a mountain out of a small molehill.

As Benoit Mandelbrot has written:

> Next, let me remind you that the new math fiasco started when a com-
> mittee of my elders, including some of my friends, all very distinguished
> and full of goodwill, figured out among themselves that it was best to start
> by teaching small kids the notions that famous professors living in the
> 1950s viewed as being fundamental, and therefore simple. They wanted
> grade schoolers to be taught the abstract idea of a set. For example, a box
> containing five nails was given a new name: it became a set of five nails.
> As it happened, hardly anyone was dying to know about five-nail sets.

> (Frame and Mandelbrot 2002)

Topics should be sufficiently rich and complex for pupils to be able to develop
their own (simple) intuitions and pose their own interesting problems.

Unfortunately, sets still appear in all mathematics syllabuses and textbooks, although the notation of sets is often inappropriate, from a professional perspective, as was pointed out during the modern maths era.

'Maths is fun'

This is the twin of the last claim. Maths can be 'fun' if you don't take it
seriously — but children tend to take everything seriously, unless they are
bored. Observe children playing football and notice what happens when
one player ignores the rules or 'fools about'.

Individual activities in the classroom may be 'fun' but mathematics as
a whole is a serious business — like art, or science, or literature. No wonder recent research showed that countries where maths is *not* regarded as
'fun' have higher achievement.

The claim that 'maths is fun' is not so much a lie, as an insult, as if
mathematics were an amusing pastime, a bit of a laugh, but nothing more.
The claims that 'maths is easy' and 'maths is fun' are especially damaging —
indeed, disastrous — to weaker pupils who know that maths is difficult
and won't do better by being insulted and lied to.

This does not mean that mathematics cannot be enjoyable. It is an
English error, a grave mistake, to suppose that 'fun' and 'enjoyable' are syno-
nyms. They are not. Maths should be presented not as fun but as enjoyable,

surprising, mysterious, challenging, powerful, amazing, and extraordinarily useful, as well even as 'neat' and 'cool', as we shall see when we discuss beauty and aesthetics in maths from the school pupil's perspective.

Steep curves of difficulty

Mathematics is not only a difficult subject, the difficulty can increase dramatically with small changes in subject matter or approach. The difficulties that so many pupils have with fractions illustrates this point perfectly. So does this account of Joshua, aged 6 years, using counters to work out how to share 16 lollies between himself and two friends. What will he do with the single lolly left over?

Joshua: Well ... You could have the extra one yourself, or you could give the extra one to your mother.

Perry: Yes, are there any other things you could do?

Joshua: Yes, you could cut it into two pieces and give each one of friends half each.

Perry: Right, but anything else you could do?

Joshua (after some thought): Yes, you could cut it into quarters and each could have a quarter.

Perry: Would that use all the lollies?

Joshua: Yes, well, really not quarters, no, they're sort of halfway between a half and a quarter.

(Perry and Dockett 2002)

Fractions and increasing difficulty

Frank was 12 years old. On being asked to add 1/2 and 1/3 he was at once stuck. However, he could halve 1/2 to get 1/4 and that to get 1/8 and so on. Indeed, he continued the sequence for a long way:

1/2 1/4 1/8 1/16 1/32 1/64 1/128 1/256 1/512 1/1024 1/2048

...

and even further, doubling in his head from successive answers.

(*Continued*)

(Continued)

He was also able to add   1/2 + 1/4 to get 3/4
1/2 + 1/4 + 1/8 to get 7/8
1/2 + 1/4 + 1/8 + 1/16 to get 15/16
1/2 + 1/4 + 1/8 + 1/16 + 1/32 to get 31/32

and so on, and then to spot — which is much easier — the pattern in the answers so that he could *predict* future answers in the sequence.

This is typical of many pupils who find 'binary fractions' which only involve only *doubling* and *halving*, much easier than general fractions.

Fortunately, pupils can learn to add, subtract, multiply and divide binary fractions without knowing anything about general fractions.

Moving from binary fractions to general fractions — even to thirds and fifths plus binary — is a really big leap in difficulty for many pupils. General fractions are *much* harder, a phenomenon akin to the *problem size paradox.* (See also the note on page 39.)

Perhaps it is a pity that Perry did not ask Joshua to draw a picture to show how the last lolly might have been divided and whether he could think of a name for this 'thing' which comes 'halfway between a half and a quarter'. Note that Joshua has, apparently, no conception of a third, but he seems okay with simple halves and a half of a half (Perry and Dockett 2002:87).

Difficulty also rises sharply, of course, when new and perhaps more abstract topics are introduced. Fractions are for most pupils, if they are to understand them really well, extremely difficult, which is ironic given that pupils start on fractions in primary school. Elementary algebra is also difficult, not least because of the leap in abstraction. These are the two biggest barriers that pupils face, but there are many other smaller but still daunting hurdles.

Even simple increases in the size of the numbers in a problem — the 'problem size' effect — can be daunting: people take longer to solve all basic arithmetical problems, such as 12 + 5 = 17 when the numbers involved are larger, and they also make more mistakes. Browne (1906) first noticed this effect in his research and it has often been confirmed since. It is related to the more general point that slight changes in language or notation or generality (abstraction!) can make large differences to the difficulty level for most pupils (Campbell 2005:325) (Zbrodoff and Logan 2005).

The curve of difficulty rises steeply, not smoothly but in jumps, and with every jump upwards a proportion of pupils are in danger of being left behind.

The steep ability curve in mathematics

It is also true, however, that there is a very steep ability curve in maths. Some pupils do find even the apparently simplest ideas very hard while others sail through such 'easy' concepts only to meet their own 'ceiling' later at a higher level.

The strongest pupils take to algebra like a duck to water, they find it perfectly obvious that if you multiply $(x + 1)$ and $(x - 9)$ you get $x^2 + x - 9x - 9$ (having previously found the idea and properties of negative numbers also perfectly clear and obvious) and they can leap ahead to solving quadratics by the most abstract means expected at secondary level.

Most pupils, however, will find them much harder, and very many pupils will find the slowest and most elementary methods of solving the puzzle quite hard enough — and if expected to go beyond that level, will at best be able to solve quadratics by what is no more than a 'rote' understanding … which does not benefit them at all.

In this case the ability they are demonstrating, say, in answering an examination question, is of little value or credit to them, while the ability they would be demonstrating by solving the problem by a more basic method would be real and of value — because they understood it.

If we observe a class over a period of time, then we find that the stronger pupils not only understand successive topics better but they 'pick up' and retain a repertoire of ideas and possibilities and tricks of technique that escape the weaker pupil who is struggling merely to keep up. Consequently, the actual difference between the strongest pupils and the weakest is far greater than it appears from test and examination results — striking though that appearance is.

Matching curves of difficulty and ability

We want to build up the curriculum on a foundation suitable for all, rather than construct a curriculum for the strong few which is then watered down for the majority. *Because the curves of difficulty and ability are both very steep, this is possible.*

All pupils can and should study a basic syllabus, sufficiently slowly to grasp the concepts more-or-less firmly. Stronger pupils will study markedly more difficult material — matching their much greater strength — and the few very strong pupils will be able to study on top of the foundation syllabus, much harder concepts matching their far greater ability.

Therefore it should not be necessary for the majority of pupils to study syllabuses that are beyond them for the sake of a minority who will go on to study mathematics at higher levels, and the argument promoted by Cockcroft and others that syllabuses should be built upwards from a foundation which suits all pupils is both morally and pedagogically sound and practicable.

The end results of mathematics teaching

Felix Klein compared a high-school pupil with a cannon which is being charged with knowledge for ten years and then fired after which nothing remains. (1997:242)

There are many difficulties in teaching mathematics and many failures: lying behind them all, however, is question that is almost never asked.

According to a BBC survey (August 2007) of more than 3,000 adults who had studied a foreign language at school (which usually means for five years) each remembered, on average, seven words, which suggests that their years at school might have been much better spent, unless they were equally unsuccessful in remembering what they 'learnt' (so badly) in other subjects also.

Coincidentally, a survey by *What Car?* magazine in the same month concluded that three quarters of active motorists would fail the driving test if they had to retake it.

These results suggest a pertinent question: what are the lasting effects of our mathematics teaching? How confident should we be that our teaching actually *has* lasting effects? How much do most pupils still grasp and retain after five years of secondary mathematics?

Hermann Ebbinghaus, the pioneer nineteenth century memory researcher, suggested (I do not have the exact words but this is the gist of his claim) that most pupils forget *most of what they learned* in school within a *few months* of leaving. More recently, Jean Piaget argued that:

> ... we are postulating that success in those examinations constitutes a proof of durability of the knowledge acquired, whereas the real problem, still in no way resolved, consists precisely in attempting to establish what remains after a lapse of several years of the knowledge whose existence has been proved once by success in those examinations, as well as in trying to determine the exact composition of whatever still subsists independently of the detailed knowledge forgotten.
>
> (Piaget 1970)

Alex Bogomolny comments on his Cut-the-Knot website:

> With all the investment in mathematics education and in mathematics education research, it is absolutely unbelievable that Piaget's query remains unanswered up to this day. I suspect that the reason is that the answer is mostly known. Establishing the fact scientifically would probably shatter the whole hierarchy of mathematics education. A math education reform should start from the cradle and continue through K-16 in a concerted manner. As long as mathematics is taught for its pragmatic value, no such reform may be seriously contemplated.
>
> (Bogomolny 2007/2015)

How true is Ebbinghaus's claim today in the UK? What research has been done on the fate of mathematics learnt at school, in pupils with different destinations? As far as I know, none, none whatsoever.

Judging from experience, I would suspect that Ebbinghaus and Klein were pretty near the mark: pupils who do not go on to use their maths in a way that reinforces and develops it (as driving a car reinforces at least some of the learner driver's lessons) will forget much of what they 'learnt' very quickly, in which case what on earth is the justification for teaching it? No doubt Alex Bogomolny is correct also: even raising the question opens up frightening vistas for the mathematics profession.

Transfer of training arguments would justify learning even if you forgot the content instantly. If you believe in the nineteenth century military metaphor that treated mathematics as a means of discipline and training, as the drillmaster of the intellectual faculties which would organise and strengthen them and bring them to peak fitness, then you will be relaxed about the possibility that pupils quickly forget all the content they have ever learnt because the positive effects will remain — but hardly anyone believes today in the simple transfer model — so what value does mathematics have for millions of children who do not use their equations, trig functions, or similar triangles after they leave school?

Much of the present syllabus is entirely unjustified for many pupils *in the form in which it is currently presented*. Many pupils, perhaps most of them, know or believe they know that they will not need most of their mathematics after leaving school — so they are demotivated from the start.

> ... mathematics is not only taught because it is useful. It should also be
> a source of delight and wonder, offering pupils intellectual excitement,
> for example in the discovery of relationships, the pursuit of rigour and
> the achievement of elegant solutions ...
>
> (DfE 1988; in Goulding 2003:128)

'It should also be', 'It *should* ...' Yes, indeed, but does it? This passage sounds like an invitation to some penetrating research. Does the current National Curriculum, as implemented via the medium of examinations syllabuses, typical textbooks and teachers loaded with targets, targets and then more targets, achieve these ideals? If not, then why not?

Bill Thurston

The late Bill Thurston was a world-famous mathematician, a winner of the Field's Medal (the mathematicians' Nobel Prize) and an acute commentator on the professional maths scene.

In the following passage, having started by damning mathematicians who turn up to give a talk at a conference and plunge immediately into the deep end where those members of the audience who attempt to follow them are soon drowning in incomprehension and only stay in the room out of politeness, Thurston turns his attention to *university classrooms*:

> This pattern is similar to what often holds in the classroom, where we go
> through the motions of saying for the record what we think the students

'ought' to learn, while the students are grappling with the more fundamental issues of learning our language and guessing at our mental
model. Books compensate by giving samples of how to solve every type
of homework problem. Professors compensate by giving homework and
tests that are much easier than the material 'covered' in the course, and
then grading the homework and tests on a scale that requires little
understanding. We assume that the problem is with the students rather
than with communication: that the students either just don't have what
it takes, or else just don't care.

(Thurston 1995:98)

Sounds familiar? Does this account perhaps tally with the experience of
some secondary school teachers? By sheer coincidence, only a few days
before writing this I had been explaining to a boy of 16 (whom I will call
Peter) who had serious problems with maths that (paraphrasing):

Teachers often realise that not all the class have understood what the lesson has been about, so they sometimes try to present the conclusion as a
'rule' which pupils can use if they haven't quite understood. Textbooks
collaborate with the teacher by presenting their own pretty clear explanation of the ideas involved — but then add a rule anyway, for the same
reason, to help pupils who haven't 'really' understood the result,

(I was now directly addressing Peter's own experience, illustrated by pointing to the very thick textbook he was using):

The result is that pupils tend to grab at the rule and forget the explanation, but they don't even get the rule perfectly, so they end up half understanding a rule which half represents what the topic is supposed to be all
about — and they get completely lost.

Peter recognised this account at once. And like a geometric sequence, the
result of half understanding this and then half understanding that and half
understanding the other is that after a few years the pupil's total understanding is not 1/8 of what it should be but 1/16, or 1/32 or even 1/1024.

No wonder that a significant percentage of pupils in their secondary schools can't even master primary school mathematics.

NOTE: Too late to incorporate in the box on page 33 I came across this quote from the Cockcroft Report (1982:75):

Although fractions are still widely used within engineering and some other craft work these are almost always fractions whose denominators are included in the sequence 2, 4, 8, ... , 64. This sequence is visibly present on rules and other measuring instruments and equivalences are apparent. Addition or subtraction of lengths which involve fractions of this kind can be done directly by making use of the gradations on the rule. When the calculation is carried out with pencil and paper it is never necessary to work out the common denominator which will be required because it is always present already; for example 2 1/4 + 3 5/16 has the necessary denominator, 16, already visible ... The need to perform operations such as 2/3 + 3/7 does not normally arise.

3

Motivating and Motivated Pupils

Philosophy

> All mathematical pedagogy, even if scarcely coherent, rests on a philosophy of mathematics.
>
> (Thom 1973:204)

This is one of the most common quotations in all of maths education, at least in the English-speaking world, and rightly so because René Thom puts his finger on a crucial point: every writer writes from a perspective and that includes some idea of what mathematics 'is' and how it 'works'. My own picture, which will be elaborated in Chs 7 to 11, is that mathematics can be most effectively seen from three perspectives:

- The objects of mathematics exist in much the same way as do the pieces (or boards) of abstract games such as chess or go, or the cards in a card game. This explains why there is no sharp distinction but a large overlap between all the abstract games, mathematical recreations, and mathematics, as illustrated by the title of Martin Gardner's long-running column in *Scientific American* magazine, 'Mathematical Games', which included everything from pure mathematics to puzzles and conundrums old and new, and recreations such as pentominoes and dissections, to abstract games such as chess, go, sprouts, hackenbush, Golomb's game and tic tac toe.

It also explains why so much mathematics, like chess, can be 'done in the head', mentally, within the limitations of memory (admittedly very severe for almost everyone).

Finally it explains why proof and an extraordinarily high degree of certainty in one's conclusions are found in abstract games (if you are talking about actual analysis of moves: our judgements are extremely fallible) and in mathematics, provided that the proof concerns concepts that are sufficiently well understood, and is not too long and complicated. (Many published proofs do contain errors, and proofs about objects ill-understood at the time, such as 'polyhedra', have often been in error too: but on the other hand, hardly anyone doubts even in their wildest moments that, using ordinary counting numbers, $13 + 17 = 30$ and that on the Euclidean plane, Pythagoras's theorem is true. At a slightly higher level, every prime of the form $4n + 1$ is the sum of two integral squares in one way only; and so on and so forth. Professionals use very large numbers of theorems in which they have complete confidence, with good reason: their proofs are convincing and they form parts of networks of propositions which hang convincingly together.)

- Mathematicians can be seen as working within miniature worlds, each of which is related to many others, each of which has its own rules, but each of which can be developed and expanded by creating and adding new ideas and concepts, new objects, new categories and so on. (So that is one striking difference from chess in which the rules, if you are playing a standard variant such as the occidental chess recognised by FIDE, change extraordinarily rarely and only, in recent decades, in minute respects.)

  These *miniature worlds* can be explored in a thoroughly scientific manner, seeking and collecting data, forming hypotheses and testing them. However, once again, mathematics differs from the natural sciences because it has recourse to proofs which not only create very high levels of certainty and confidence, far beyond those available to the natural scientist. They also by depending for their creation on new ideas and concepts, 'force' mathematicians to understand their subject more and more deeply. The best proofs create not only confidence in the result, but illumination.

- These two aspects of mathematics, the game-like and the scientific, both depend on perception: mathematics is about 'seeing', where those quotation marks indicate that this is often not — though it often is — about literally seeing a geometrical figures and its features, or even seeing the features of an algebraic expression (such as its symmetry, or the order of the exponents). 'Seeing' can be highly abstract, and indeed many pupils find that the 'seeing' demanded of them in school lessons is very abstract indeed and therefore very difficult.

Extrinsic and intrinsic motivation

Motivation, to make a rather crude distinction, comes in two forms, extrinsic and intrinsic: we cannot rely on the former in this country because learning and education are not sufficiently respected, some subcultures apart. (Many cultures are more fortunate in this respect as we shall note in Ch. 12.)

Examinations are not a sufficient motivation to keep most pupils going over a five to ten year period and for those many pupils who start to fail early all tests, far from being a positive motivation, are a source of anxiety and dread.

Extrinsic motivation in this society being as feeble as it is, it is therefore even more important that intrinsic motivation be as powerful as possible.

Motivating pupils through the pupil's own activity

I do believe that problems are the heart of mathematics.

(Paul Halmos 1980)

Paul Halmos had a dynamic perspective on mathematics emphasising the *personal activity* of the mathematician. We might add, quoting the *Plowden Report* (1967): 'At the heart of the educational process lies the child,' though at the risk of being satirical — we know what Plowden meant — we might add that the child should actually be jumping about, climbing and stretching and being extremely active, *mentally as well as physically*, recalling the Nuffield Mathematics Project motto, 'I do and I understand'.

Solving problems is an activity of a different order from 'simply' learning concepts. Its social, emotional and cognitive connotations are stronger and more meaningful.

Children of all ages are most highly motivated when they are most deeply involved in their own activity, with others or by themselves. (So are adults.) Pupils learn by doing, which includes thinking, and not just by listening — though you hope that they are thinking *while* they are listening.

These *doing* activities include admiring and making pretty patterns, constructing objects, constructing solutions to problems, thinking of good arguments and constructing proofs, doing experiments and making observations, constructing hypotheses and testing them, and so on.

Such activities, as all primary teachers know but secondary teachers sometimes forget, are for all pupils, of all abilities.

Ken Hutchon of the SCEEB was a field officer for Foundation Maths, developing maths courses and assessment procedures for the least able children. He remarked, in *The Problem Solver* #3, 'Editorial', 'how an incredible variety of people have the persistence and logic and imagination to try to solve really difficult problems if the problem grabs their attention in the first place'.

He enclosed examples of problems presented to 'bottom 30%' pupils in a trial 'Mathematics Extended Response Test'. The report by the examiners included comments such as, 'This question was very well done and most of the markers … have been very surprised (even astonished) by how well pupils coped …' 'An amazing number of pupils scored full marks …' (1980–1983, Newsletter #3:4).

It has been a mark of British maths education in recent decades that every so often someone comes with just such novel opportunities for children — which then fail to take root and wither and die.

Maximum motivation

A very old and reliable 'law' in psychology, the Yerkes–Dobbs law, says that for maximum performance anxiety should be neither too high nor too low, but somewhere in between. We might apply the same idea to motivation (so often linked to anxiety) and conclude that for maximum motivation the task will be neither too difficult nor too easy.

In other words, as the work becomes *easier*, so motivation tends to go down. The trick therefore is to keep the work demanding and 'hard' but also at a cognitive level that the pupils can master: which means giving them time. It also means that the teacher has to think backwards, and downwards, to the thought processes of the pupil.

When is a problem solved?

It is natural to suppose that a problem has a solution, and that when the solution has been found, that is an end of the matter. This naive view, however, is contradicted by the behaviour of professionals, and by recent tendencies in mathematics education in this country to emphasise problem solving and investigations, in which pupils explore a problem and record their discoveries. Exploration may indeed lead to a solution, but it may also lead to new ideas, new problems and new goals in a way that undermines the traditional assumptions.

A similar tendency can be seen in the long history of the problem section of the *American Mathematical Monthly*, a magazine aimed at a college audience. The *AMM* started in 1894 as a journal 'devoted to the solution of problems in pure and applied mathematics'. The early problems and solutions published much resembled textbook exercises.

In 1932 the problems were split into elementary and advanced problems. The latter section claimed to '(especially) seek problems containing results believed to be new, or extensions of old results ...' There is a hint here that a 'result' might not be the end of the problem, though in practice solutions were submitted and published as before.

At the end of the 1960s a new section started, called 'Research Problems'. Richard Guy edited it from 1970 onwards and in odd years published updates describing progress on the published problems. Over a period of roughly a century, the idea of problems had changed from being no more than a textbook exercise, to being, potentially, a challenge, open to development.

The idea that a problem leads to *developments* is a commonplace among professionals. Hilbert's 23 problems of 1900 took for granted that the solution of such difficult problems would lead to important advances in mathematics, and they did (1993b:25).

Puzzles, problems and exercises

Fortunately, puzzles have an almost universal appeal across time and space, provided (of course) that they are presented in a form that is attractive to the potential solver. Fortunately, also, so many puzzles have a strong mathematical aspect as do ('by definition') mathematical recreations. The

Combinatorial puzzles

In how many ways can four pupils be lined up in a straight line to face the camera?

In how many ways can a class of 30 pupils line up in a straight line to face the camera?

Is it possible to arrange nine points on nine straight lines so that there are three lines through every point and three points on every line?

The farmer wishes to catch the pig. The farmer moves one square up, or down, or left or right at each turn. Then the pig does the same. The pig is caught when the farmer moves onto the square on which the pig is standing. Can the farmer catch the pig?

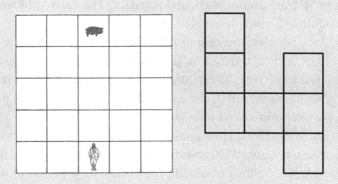

The farmer chases the pig   A tiling for the plane?

Is it possible to tile the plane with repeated copies of this tile, which can be rotated or turned over?

If five points are marked in a unit square, then at least two of them are less than 0.71 units apart.

(1987a:Unit 27: Combinations)

True or false?

'If there are six people in a room, then there are 3 who either all know each other, or none of them know each other.'

combination of simplicity in the posing of such puzzles, and the unsuspected depths to which they lead, means that a wide variety of pupils can tackle them at a suitable level without the possibilities in any of them being exhausted.

Young children are especially attracted by puzzles and primary teachers know this but somehow as pupils get older the message becomes garbled:

> Yes, puzzles are motivating but they are kind of non-serious, they are just a bit of fun and mathematics is a serious subject (except when I'm trying to motivate my pupils by claiming that 'maths is fun') and anyway solving puzzles takes a lot of time and some pupils fail, and I have a syllabus to plough through, so I can't really rely on puzzles — except for a bit of fun — oh! I've said that already! — and so I'm going to teach them simultaneous equations from the front of the class, and when they've done the exercise — if they get to the end — then the stronger ones can do some problems, yes, that's it, teach, exercise, problems for the better ones, and what about the weak ones? Well, they'll try with the exercise, as usual, won't get very far. Too bad! What do you expect?

'Not enough!' is the right answer. Puzzles and problems are a challenge that offer an intrinsic reward. G.H. Hardy referred to the 'kick' that so many people get out of mathematics:

> We may learn the same lesson ... from the puzzle columns of the popular newspapers. Nearly all their immense popularity is a tribute to the drawing power of rudimentary mathematics, and the better makers of puzzles ... use very little else. They know their business; what the public wants is a little intellectual 'kick', and nothing else has quite the kick of mathematics.
>
> (Hardy 1941:86–88)

An inevitable concomitant of giving pupils choice and allowing them to express their opinions of particular problems is they may decide that they do not like what is offered, that they do not want to tackle the problem suggested. This raises the affective level but also raises complex practical and moral problems.

Of course, if pupils are given an overview of their course, *and if they accept this picture*, then your problem is solved.

Problems, puzzles and motivation

> Those who study mathematics cannot begin too early the exercise of their talents with the solution of problems presented by that science; for it is by such exercise that the inventive faculty is called forth and strengthened. We have therefore thought it our duty to subjoin to this part of the Mathematical Recreations, a selection of problems proper for exercising and amusing young mathematicians.
>
> (Hutton 1840:183)

Puzzling usually means intriguing and attractive (to most people — some more than others) and therefore motivating, though it can also have the connotation of triviality and lack of seriousness. In this book I shall assume that puzzles are serious, and challenging. Unfortunately, most standard textbook and syllabus topics are not turned into puzzles.

At the other end of this spectrum, exercises are usually little more than tests of the pupil's ability to reproduce what he or she has just been taught. They may be highly motivating to pupils who love the idea of high marks and pleasing the teacher but they usually have no intrinsic motivation and far too many pupils 'solve' them by using the rules which the teacher has, in effect, taught them.

For pupils who understand the teacher (and topic) well, exercises may serve to fix ideas in the memory. Unfortunately, for many pupils who did not understand the teacher clearly, they fix misunderstanding, or weak understanding and create a sense of failure.

In between comes the problem: this could be close to an exercise, requiring only routine application of ideas the pupil is supposed to have grasped, or it could be close to a puzzle with a strong original element that forces pupils to think for themselves.

Changing expectations

If pupils are to do as well as possible at solving puzzles and problems, they need appropriate expectations: here are some of them.

Pupils may be given a choice of problems, inviting them to exercise their own preferences and giving them a sense of responsibility for tackling the problem they choose. In parallel with this choice, however, they should be allowed to abandon a problem if they decide after trying it that is not what they thought, that they do not like it after all: this is what professionals often do. On the other hand, they should not make such switches lightly.

They need to learn through a mixture of their own experience and commentary from the teacher — who is effectively their *coach* — that some problems are more enjoyable than others, that there is more than one way to solve a problem, they will often get stuck (as professionals do), that they will sometimes fail completely and that you cannot solve every problem — and neither can the teacher! Also, there is much luck in problem solving.

The pupils also need to learn that their own ideas are valuable, and that their own questions may themselves be interesting problems. Also that there is no rush, and that success is to the pupil's credit (or the pupils' credit, because working together with other pupils is good, and not cheating and pupils learn from and teach each other.)

Solutions to puzzles should also be seen as not merely right or wrong. The answer 'Yes!' is just that, an answer not a solution. A superior solution includes an explanation, reasoning, and maybe commentary on the ideas that led to it, or on further questions that it suggests. It follows that justifying your solution is a problem in itself, in English as well as mathematics (1987a:5–8).

Choosing your own level of difficulty

Mathematics is difficult for everyone not least because professional mathematicians choose goals that are commensurate with (what they believe to be) their talent. They aim high but not absurdly high. When Andrew Wiles chose Fermat's Last Theorem — how ambitious! — he was a brilliant student at Cambridge who was expected to go far. We might say that his judgement of his own ability was spot on: after the proverbial seven years of struggle, his first 'proof' contained a gap, which he filled with the help of his student Richard Taylor but only after hovering for a year or more between dazzling success and gut-wrenching failure (and the loss of the post of head of mathematics at Princeton University which Wiles now occupies).

There is no point in a professional being satisfied with 'easy' maths: that way you do not get professional advancement and you get little or no personal satisfaction either, so professional mathematicians resemble rock climbers and mountaineers who reject shallow slopes and large hills because they are too easy, but also avoid the north face of the Eiger or the peaks of the Himalayas unless they believe that their talents and experience are up to the challenge.

In sports generally, individuals and teams play against those of much the same ability not only because mismatches are unfair but because they are not enjoyable for the players, or the spectators. Experience shows also that the most beautiful games come from struggles between more or less evenly matched opponents.

One consequence is that professional mathematicians often have to study their colleagues' unfamiliar papers very hard indeed, following the detailed argument not just page by page but line by line.

(Published papers are notorious for often not explaining clearly in advance what they are doing: they tend to be dense and obscure, so that writers such as Euler who explained what they were doing and whose papers are therefore easy to read, are famous for that fact.)

Also, simply following another person's argument often results in only a low level step-by-step understanding, so professionals often try to recreate the argument themselves not because they want to waste time but because they want to develop a deeper and *intuitive* understanding of the original. Even at the highest professional level, *doing the maths yourself* is preferable to merely following the mathematics as done by someone else!

The difference between professionals and school pupils is that the pupils are usually given no choice in what they tackle. Today is Monday, so it is simultaneous equations from Book 3, Ch. 6 and if some pupils find this hopelessly difficult while others find it dead easy, too bad! That's the way of the (scholastic) world, and bad for motivation.

In contrast, when pupils are given a topic and choice of problems to tackle, they can be very highly motivated indeed.

Whatever the level of difficulty, observation of learners of abstract games, or of a musical instrument or any other skill, suggests another important phenomenon. Pupils develop not steadily but in jumps, resting in between on a plateau on which they may be 'stuck' for a long time before suddenly, and inexplicably, jumping a level. Contrast the assumption that pupils ought to learn at the rate at which they are taught by the teacher.

Taking things slowly: turning a steep cliff into a gentle slope

> Research as well as psychological and logical scrutiny made it unmistakably clear that skills which to the adult are exceedingly simple are to the child far from simple. We had been guilty of asking children to progress by jumps or steps that were too large and that were presented too rapidly.
>
> (Brownell 1937:3)

We considered earlier how the difficulty of mathematical topics can increase very rapidly. The typical textbook rises steeply upwards to a high level of abstraction and difficult explanations, but this curve can always be flattened out and turned into a longer and gentler slope: the result is that the topic becomes not only more accessible to pupils and more motivating, but also richer for them.

Problems can be made easier or more difficult by the simplest means, often just a small change in data or wording, or by taking an easy case first and leaving harder cases until later.

The cover of *The Problem Solver* #7 showed this figure:

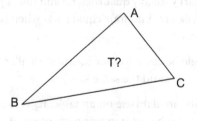

Can three areas be equal?

It then asked: 'Can you find a point T, somewhere inside this triangle, so that the area of the three triangles ATB, BTC and CTA are equal?' Such problems suggest many variations; some easier, some harder. In all these puzzles or problems, two points stand out. One is that the difficulty of the various methods of solution, techniques and approaches varies enormously; the second is that the more abstract the solution (meaning, often, the closer to standard technique) the harder it is for pupils to understand.

If pupils are doing an exercise in which they are demonstrating their understanding of the teacher's previous explanation, then inserting much harder exercises may floor even stronger pupils if they automatically expect the solution to be found *via* the teacher's explanation rather than by thinking anew for themselves: but if pupils are solving problems using their own initiative and ideas, then stronger pupils will easily be able to tackle harder problems, so we might say that problem solving is a vastly more flexible and adaptable form of learning than instruction from the

Varying the problem

1   How can a triangle, any triangle, be divided into two parts of equal area by one straight line? How many solutions has this problem?

3   How can a triangle be divided into three equal parts by two straight lines? Generalise!

4   What is the smallest number of lines needed to divide a rectangle into 12 equal parts? Ten equal parts?

5   How many planes are needed to divide a rectangular box into eight equal parts? 20 equal parts?

7   Is it possible to start with any quadrilateral and find a point inside it which divides the quadrilateral into four equal parts when the point is joined to each of the vertices?

9   How can a triangle be cut into four identical smaller triangles? (Identical means the same shape and the same size.)

12   A cone rests on its circular base on an table. Its vertex is above the centre of its base. How can it be cut into two parts of equal volume by one plane slice parallel to its base?

16   Is it possible or impossible to divide a rectangle into five triangles of equal area?

(1980–1983, Newsletter #7:11–12)

teacher. Here is an illustration of change-of-language which though strik-ing is much more than a caricature:

Version 1: 'The interior of a simple plane curve is divided into a finite number of simply connected regions, each of which is assigned to one of the classes R_i (i = 1, ..., n). If no two regions bounded by the same edge are assigned to the same class, and if the number of classes R_i is a mini-mum, find the value of n.'

Version 2: 'What is the smallest number of colours needed to colour a plane map, if any two adjacent regions are of a different colour?'

The first version uses many mathematical terms and ideas and sounds dif-ficult for that reason alone. The second version uses everyday terms, apart possibly from 'plane', and could be understood by almost anyone. Significantly, mathematicians themselves usually use Version 2 rather than anything resembling Version 1.

Dynamic versus static

Nothing is more natural (and motivating) for an applied mathematician or a child than to think of functions behaving, doing things, reaching a maximum, tending to a limit, rushing off to infinity and so on. The idea of *movement* is one of the great metaphors that mathematics takes from real life and which small children meet the moment they start to count. So it is no surprise that the *Association for the Reform of Geometrical Teaching* (the forerunner of the Mathematical Association) decided in 1871 that,

> In any new textbook ... the following principles, only partially or not at all recognised by Euclid, should be adopted ... (iii) Superposition. (iv) The conception of a moving point, and of a revolving line.
>
> (*Association for the Reform of Geometrical teaching* 1871:2)

Here is professional mathematician Stan Wagon talking dynamically about the Riemann zeta function (the famous hypothesis that remains unsettled after more than a century) and its behaviour, using vivid imagery that has an immediate appeal:

> Certain patterns in the zeta function have been discovered ... Exceptions
> to these patterns exist and although a fatal exception has not been found,
> there have been some near-misses that show how fragile the situation is.
> It seems that even a slight breakdown in the relatively uniform behaviour
> of the zeta function could lead to a counterexample.

(Wagon 1986)

Jean Dieudonné was a prominent member of the Bourbaki group and a
proponent of the New Maths which emphasised the static language of sets
and structure, but here even he uses dynamic language: readers can ignore
the technical terms and simply note the word *misbehave*:

> Homological algebra ... started in fact as a kind of glorified linear alge-
> bra, by introducing concepts ... which in a way measure the manner in
> which modules over general rings *misbehave* when compared to nice
> vector spaces ...

(Dieudonné 1964:244)

All of a sudden, abstract mathematical structures are misbehaving like
naughty school pupils! Mark Kac showed a more self-conscious use of an
even more vivid metaphor:

> This holds, of course, for any number of primes, and we can say, using a
> picturesque, but not very precise language, that the primes play a game
> of chance! That simple, nearly trivial observation is the beginning of a
> new development which links ... number theory ... and probability
> theory.

(Kac 1959)

The next quote suggests that numbers may be almost physically hidden
from view:

> If one is focusing on a particular sequence of positive integers and one
> wishes to sense the presence of the integers without actually counting

them, one ... studies some other phenomenon that is sensitive to their existence. We refer to this as 'indirect counting'.

(Shapiro 1983)

Note how similar this strategy is to that of physicists who cannot, for example, observe molecules directly so instead observe their effects in a Wilson cloud chamber.

Descartes conceived of his original co-ordinate geometry in terms of a moving ordinate line. The familiar co-ordinate plane is actually a static picture — just as a geographical map is static — of this original dynamic scenario, perfectly illustrating how the passive and active modes are both invaluable.

From primary school onwards there is a dualism between possible interpretations, active and dynamic or passive and static. Sometimes the static is easier for pupils to appreciate, sometimes the dynamic. The poet Coleridge noticed that according to Locke, four and five *are* nine, but Coleridge himself thought that they would *make* nine (Coleridge 1905:145). We can say both are correct — it depends on your perspective, active or passive.

Pure mathematicians have good reasons for replacing the dynamic imagery so often used by their great predecessors, such as Fermat, Descartes and Newton, with static alternatives — but their reasons do not apply to school children for whom a computer which shows a curve being plotted by a moving point in real time can be much more vivid and motivating than an abstract set of points.

Surreptitious learning in rich environments

There is another way in which problem solving saves time in the long run. Because of the richness and variety of problem solving, pupils are certain to meet ideas which are some way away from the ostensible subject of their problem. In so doing, they will make connections and develop ideas which will be of great use to them later. An example is concepts of graphical representation which pupils develop from the simplest charts and pictographs and game boards from primary school onwards. There should be NO occasion when pupils meet the idea of a graph or graphical representation for the first time.

Pupils will also, unknown to themselves, be developing more subtle understandings that are more difficult to put into words. A traditional tendency assumes that all that the teacher needs to convey to the pupils can be put rather simply into the English language, indeed into rather straightforward explanations, and that this will transfer to the pupil the substance of the teacher's own understanding. Neither the teacher's nor the pupils' understandings are so simple. Professional mathematicians know full well that it is not enough in order to enter a new area, simply to learn the main results and methods from a book. It is necessary to work in the field, to get a feel for it and to develop their intuition by solving problems and being mathematically active. Children need just as much to develop their own feelings and intuitions and subtle ideas through their own mathematical activity (1987a:22).

Negative numbers are a topic that should be introduced surreptitiously and bit by bit over a long period of time. For example, negative numbers as used in everyday life and in science for negative temperatures or for backwards movement can be added and subtracted many years before the multiplication and division of negative numbers in encountered (and those operations might be encountered via questions which arise naturally when negatives *are* added and subtracted).

Pupils should first meet new concepts casually. They should be first introduced surreptitiously, so that pupils who are likely to find them tricky can have several bites at the apple of understanding without the likelihood of demotivating failure. Surreptitious learning is marked by the fact that children cannot 'fail' because nothing is being explained — they are *not being taught* — and there is therefore nothing that they can explicitly fail to understand. Children spend so much time in school failing because they spend so much time being taught. In real life they spend more time learning, and fail less.

Explanation can lead to the level of misunderstanding being higher at the end of the lesson than it was at the start. Struggling pupils may start off with a certain level of misunderstanding and end up with an even higher level.

Surreptitious learning is extremely common in real life, indeed, we might say that it is the norm. A child learns a vast amount about driving a car, even when small, so that learning later to drive and pass the test are much simpler than they would be to a genuine novice.

Surreptitious learning gives pupils more time, because they can be given repeated opportunities to learn the same idea, surreptitiously.

Problem solving is the ideal vehicle for the casual and surreptitious introduction of new ideas: obvious examples are the use of puzzles about maps and journeys to *insinuate* ideas of graphs.

Surreptitious learning is also the natural by-product of exploring any miniature world. All rich explorations developed intuitive understanding often surreptitiously.

'Slow learning'

We have been talking about *slow* learning which is also deep learning. In contrast we can describe a typical textbook treatment of, say, quadratics, as *fast* and *shallow* learning. The equation is introduced in a more-or-less abstract form, pupils are soon drawing the graph — using negative numbers and fractional values — then comes factorisation, with lots of exercises, and finally completing-the-square. The learning curve has been steep and most pupils will have had no chance to develop any strong intuitive feeling for what they are doing and will probably forget it very quickly. This *fast* and *shallow* learning is no basis for any course. What they need is *slow* learning which starts surreptitiously, exploits the richness of even the simplest ideas, poses many simple puzzles which they can solve and which leads them slowly but also much more surely to a modest but solid understanding — while the stronger pupils, matching their steep ability curve to the steep difficulty curve of more advanced ideas, are also reaching their own higher potential.

Just as the Slow Food movement is spreading (very slowly!) across Europe, if not yet the world, so mathematics teachers and pupils need a 'slow learning' movement, not for the sake of learning slowly in itself, but for the sake of learning better: we might say that 'slow learning' leads to *better digestion* of mathematical concepts and therefore to much higher long-term success, and motivation.

Of course, if there is no time for 'slow well-digested learning', too bad — but the consequences will be faster, so-called 'learning' which in many pupils results in first indigestion and then the loss of all nutrients because they were never digested properly in the first place, followed by chronic loss of appetite.

Slow learning for Peter, another failing pupil

Peter was 13. He spent many hours solving equations such as these: $2N + 1 = 5$, $3N + 4 = 16$, $2N + 1 = 8$, $5N + 7 = 27$, $8N - 2 = 30$.

He was introduced to them as think-of-a-number problems and several examples of the first two types only were presented orally. He was given no method of solving them, but found the solutions by trial and error. He was then told how the written equation corresponded to the oral explanation (like a code) and he started on the worksheet, still with no explanation of method. At first he could only guess. Then he started spotting simple patterns relating the answers to the coefficients of the problems, but as soon as he went to the start of the next type the pattern, of course, disappeared. Several of his patterns were wrong, anyway, which did not help. They explained only a few solutions — but it is significant that he was spotting patterns without being told to.

All the examples I have quoted were distinct types to Peter, each requiring a new start. Only after he had solved several dozens of examples did he begin to see that there might we some reason behind the solutions he was finding, and begin to articulate any explanation of how the solution related to the coefficients. He still did not 'see' at all clearly, and until he could 'see' clearly, the types would not coalesce into one general type.

This method was slow, but for Peter it had several good points. Peter could readily accept that this was real mathematics. It wasn't wrapped up in a surgery pill and it wasn't made babyish, so he didn't think he was being fobbed off.

He was told from the start that it was difficult, and he accepted this. He was not being patronised with easy-peasy sums (which perhaps he could not do.)

It wasn't fun, but he did not expect it to be. He thought that mathematics was difficult and that he couldn't do it. He was learning that he could do it. Moreover, he was teaching himself, and he realised that he was understanding them a bit better as time went by.

Because he was given no method to understand, he could not fail to understand the method ... (1981b)

Meanwhile, the wealth of questions that can be asked and puzzles posed, means that the weakest and the strongest pupils have plenty to think about: if some pupils are strong enough to indeed find this slow method 'tedious' then there are plenty of questions to stretch them.

Meaningful learning, understanding, and intuition

Hands up anyone who thinks that rote learning is better than learning with understanding! Aha! Just as I thought ... Traditional teaching focused on learning *how* to perform certain operations and not on understanding why the taught method worked. Throughout the twentieth century, however, there was an increasing emphasis on understanding 'why' and not just 'how'. Brownell's paper of 1937, 'The Revolution in Arithmetic', was just one marker in this progress towards so-called 'meaningful' learning.

Unfortunately, 'meaningful' learning can pose severe problems for the pupil and for the teacher: it is almost always much harder to grasp the reason behind a process than it is to simply learn to go through the process by rote. The former requires conceptual understanding and maybe new and difficult ideas, while the latter may be no more taxing than learning to operate a microwave oven. (How many people who can easily operate a microwave understand what is going on inside?)

There has been a massive failure by mathematics teachers to overcome two difficulties raised by this revolution:

- First, that teaching-for-meaning is far harder and takes far longer than simply teaching by rote.
- Second, that failure to understand is far more damaging than failure to remember.

The latter can be put down to a bad memory or forgetfulness or simply the fact that you haven't used what you learnt. Failure of memory does not automatically equate with stupidity. Failure to understand is often assumed to imply lack of intelligence, *alias* stupidity. Too many pupils learn in this way from a young age that they are more-or-less stupid, and the damage stays with them for ever.

Who is to blame for this failure? Objectively, the system first and those who have created it over decades; then the teachers who try to put it into practice; but not the pupils who have go to mathematics lessons whether they like it or not and who are then blamed when they fail to succeed at an obstacle course not of their choosing and not of their design.

Problems of motivation enter here too: learning with understanding can be very motivating while rote learning can be tedious and mind-numbing — but sometimes it's the other way round: learning with real understanding can place a tremendous strain on the pupil's cognitive capacity and so be demotivating while rote learning can be productive (because it enables you to do something worthwhile) and very easy.

An example of the latter is learning to use, by rote because the inside workings are a mystery, an electronic calculator.

At the very least the claim that learning with understanding must be superior to rote learning is shallow and naive. Our criterion should be: understanding HOW is OK if the understanding WHY could be taught but need not be because it has no further use to the pupil, now or in the future. Understanding WHY (from some perspective and at a suitable level) is necessary if the understanding will be needed in the future.

Let's illustrate this point with an example from *The Self Instructor, or Young Man's Best Companion* published in the early nineteenth century. It starts a chapter on arithmetic by presenting a very strong motivation:

> After Writing, the next necessary step towards qualifying a person for business is ARITHMETIC, a knowledge so necessary for all parts of life and business, that scarce anything is done without it.
>
> (Bowles c.1807:37)

It then describes a large number of techniques, by rote, with only the slightest attempts to conveying understanding. Here is one typical paragraph on a tricky point of multiplication:

> When you have a cipher or ciphers in the multiplier at the beginning towards the right hand, then set it, or them, backwards from the place of units towards the right hand; and when you have multiplied by the figure or figures, annex the cipher or ciphers, as in these examples. (*Ibid.*:53)

If young Frederick Willoughby has a talent for arithmetic he may easily work out the why and wherefore of this advice but if he does not then he can learn to use it in business anyway and be successful.

Later in the same lengthy chapter, following the Golden Rule or Rule of Three (on proportion), the Rule of Three Inverse, the Double Rule of

Three Reverse, and a very large number of other topics, the author reaches the subject of The Square Root where the Young Man is instructed as follows:

> Let the square root of the number 45796 be required.
>
> Set a point over the place of the units, thus, 45796, and so successively over every second figure towards the left hand, as thus, 45796. But in decimals you must point from the place of units towards the right hand, omitting one place, as above; and if the place of decimals are odd, affix a cipher towards the right hand of them, to make them even. Your number thus prepared, draw a crooked line on the right of the number, as in Division; and, indeed, the operation of the square root is not much unlike Division; only there the divisor is fixed, and in the square root we are to find a new one for each operation. Having made a crooked line, thus 45796(seek in the foregoing table [of integers 1 to 9 and their squares] for the nearest square to the first point on the left hand, which here is 4, the root of which is 2, which root place on the right hand of the crooked line, and set its square 4 under the same point, as below ...

And so on, for another one-and-a-half pages, leading to the conclusion that the required square root is 214. The ambitious Young Man could then go on to learn how to find cube roots also, by a similar process.

Once again, if Mr F.W. cannot see why the method works (which is harder than understanding the 'algorithm' as explained, lengthy though the explanation is) well, he can still use it successfully, and nothing is lost because the assumption is that he will never become a professional mathematician or even a teacher of mathematics.

What can we learn from these two examples? The first seems to our modern eyes a feeble example of rote learning, indeed of the worst sort. Instead of explaining the role of the zero or zeros, the author treats the reader like an intelligent parrot, or a hog to be led by a ring through its nose.

The second method, however, although undoubtedly learnt by rote by vast numbers of eighteenth and nineteenth century pupils — it even lasted up to the Second World War — is not quite so damnably wicked, is it? Because our pupils also use a rote method — it's called a calculator — to find square roots, with no understanding at all of how it works. The fact is that almost all pupils today use calculators for square roots, cube roots, sine and cosine functions and much more, with no insight into where the results come from — just as I and my fellow pupils had no idea where the logs and trig functions in our books of printed tables originated. So where has the demand for meaningful learning and deep understanding gone?

Oliver Heaviside (1850–1925) was a self-taught electrical engineer and a talented mathematician who defended himself confidently when he was accused of using methods of his own devising that were not at all rigorous: 'Mathematics is an experimental science, and definitions do not come first, but later on,' a point that pure mathematicians, mindful of the history of limits of sequences, or of the Euler formula for a polyhedron, can also appreciate. Speaking at the 1901 meeting of the *British Association for the Advancement of Science*, he had this forceful point to make:

> (the pupils) have also the power of learning to work processes, long before their brains have acquired the power of understanding (more or less) the scholastic logic of what they are doing ... Now, the prevalent idea of mathematical works is that you must understand the reason first, before you proceed to practice. This is fudge and fiddlesticks ... I know mathematical processes, that I have used with success for a very long time, of which neither I nor anyone else understands the scholastic logic. I have grown into them, and so understand them that way.
>
> (Griffiths and Howson 1974:18)

This is reminiscent of von Neumann's answer to a student who claimed that he did not quite understand one of von Neumann's lectures: 'You don't understand mathematics, you just get used to it!' Nonsense, of course, but ingenious nonsense with an important grain of truth hidden inside it. We can understand square roots by using them often in various situations, even though we calculate them electronically: and we can certainly say that experience will help some pupils to understand better — and later — the various algorithms and approximate methods that can be used to calculate roots 'by hand'.

The same pupils may eventually be introduced to the idea that trig functions, for example, can be calculated as the sums of infinite series. It is both surprising and mysterious to see how, by adding up the series of non-oscillating terms, the resulting curves oscillate. However, we would not teach young pupils sine and cosine functions by this method.

Understanding *vis à vis* practice

The beginning tennis player is advised to knock the ball around, to get a feel for the racket and the ball, before they receive instruction. With no instruction they are unlikely to progress far, though tennis has the advantage, which we have already noted, that it is a highly visible and public activity (unlike invisible mathematics) and so players will learn — surreptitiously, almost subliminally — by watching tennis matches at a local club or on TV.

Meaningful and enjoyable practice, and coaching (formal or informal) go hand in hand, and this parallel learning is found in almost all skilful activities — except mathematics. As Benchara Branford pointed out:

> In the days of the Hanseatic League, arithmetic was often taught to youths in mercantile offices explicitly for mercantile purposes, so that the examples and applications were not only familiar to them, but had direct bearing on the future occupation of the student. These examples and applications still persist largely in our arithmetic books and lessons, in the shape of curious tables, mercantile problems, and the like. But it is seldom seen that the justification for their existence has largely vanished: *as is so often the case, the letter remains but the spirit has gone.*
>
> (Branford 1908:267)

Many readers of *The Young Man's Best Companion* may have had the same advantage of 'learning on the job'. Ironically, with the advent of computers pupils at all levels should be able to take part in simulated businesses, but few as yet do. Primary schools do often have a 'shop' so pupils can buy and sell, practising what the teacher has been preaching but secondary schools have no equivalent.

Levels of understanding and applications

In applications of mathematics it is important to develop an intuitive understanding of the process from the applied perspective, but almost never from a pure mathematics perspective.

Given a choice between solving the pure mathematical problem involved by using a computer, or understanding the process step by step, or having a deep intuitive understanding of the pure process, the applied mathema- tician may well choose a computer or step-by-step solution plus a deep intuitive understanding of the applied *situation*.

The physicist or chemist, the geologist or biologist, the statistician and operations researcher are seldom interested in the minutiae of why some operation on an infinite series or an effective method of solving differen- tial equations works — they just want to use the method with confidence. Their own personal insights and understanding and intuition are focused on their speciality — not on the specialism of pure mathematics.

Of course, most pupils leaving school are much closer in that respect to the applied mathematician than to the pure specialist, simply because they expect to use whatever maths they may retain in everyday life and in their work. Pupils need the level and quality of understanding that is suitable to the situation they face. Professionals are in the same boat: an algebraic topologist might suspect that he needs some technique from a different area of mathematics in which he is not an expert so he asks a colleague who is, and whose understanding is deeper than his own. One day his colleague might return the favour.

Likewise, an applied mathematician designing bridges may have a deep understanding of differential equations *from his own perspective and for his own needs*, while having no knowledge or interest in the work of a pure colleague whose own researches are of no immediate relevance to bridge design.

Returning to *The Young Man's Best Companion* our conclusion should be that the old arithmetical methods of finding square and cube roots have today no further use to the pupil, and therefore using a calculator is justi- fied and no loss to the student. Likewise, the existence of series for calcu- lating the trig and other functions are of no further use for many pupils, and do not *need* to be learnt (though they make a surprising and mysteri- ous subject for exploration).

However, for stronger pupils who will go further, concepts of series soon become crucial and so their understanding — 'why' — ought to be much deeper. When we foresee that pupils will need such deeper under- standing, and with good reason, then we teach it, preferably through the pupils' own activities. For those pupils who will need a greater depth of

understanding, understanding should be available because they will —
thanks to the steep curves of difficulty and ability illustrated earlier — be
able to reach it.

Intuition

> Only a relatively small part of the total mental activity rises into the
> region of consciousness. Consciousness is merely the surface of a vast
> ocean of subconscious mental activity. Of this conscious part again only
> a relatively small part (great as that part may absolutely be) is expressible
> in language. Most thinking is done without language. The teacher cannot
> realize too clearly the profound truth that the substance, contents, matter
> (call it what one will) of ideas can never be created in the pupil's mind
> by mere spoken words, however carefully chosen.
>
> (Branford 1908:118, 119, 120, 121)[†]

We take intuition to be that level of understanding which is difficult or
even impossible to put into words. In chess or go it appears when a player
'feels' (the very word that players use) that such and such a move is right,
best, essential, or bound to work although a complete move-by-move
analysis hasn't been made and may indeed be impracticable.

Intuition gives the learner confidence that they understand a topic.
The pupil with a well-developed intuition 'feels strongly' that such-and-
such is the case even if they cannot at once explain why in so many words.
Needless to say, this is a level or type of understanding that is not present
in most secondary pupils, at GCSE or A-level.

Intuition is essential if pupils are to get beyond the stage of merely
imitating the teacher, and half-rote learning. However, intuition only
develops with experience, and experience at the right level. It is a trite
observation, but a very curious one, that however clearly an experienced
teacher explains a topic to a class, the pupils never understand the topic as

---

[†] Given the date (*Ibid.*:1908) this reads like a reference to Freud and his ideas of unconscious
activity: however, the idea of the unconscious or subconscious is much older than Freud.

The chess player's intuition

The chess player's intuition may tell him or her that Qh4 must be a good move — or that the position after the exchange of rooks is better for white. Such judgements are often ascribed at the time to *feeling* although they may be later justified by logical analysis. Such intuitive judgements are often made extremely quickly, and yet prove correct (while sometimes being grossly mistaken). The better the player, the better the intuition, which is why strong players can play extremely well, extremely fast, without relying only on rapid calculation.

> Strategy is not mere knowledge of the moves, nor is it knowledge of a few relationships between moves, or significant positions, though it includes that. Rather, strategy can be interpreted as the player's entire knowledge of the game-as-played ... An experienced player literally 'sees' the board differently from a weaker player. The same inverted image may form on the back of his eye, almost identical information may be transmitted by their rods and cones, but soon after, maybe very soon, their interpretations differ, long before the results are available to the players' conscious minds. To use a homely metaphor, the beginner finds the chess-world strange, the stronger player is well-acquainted with it, and the master is thoroughly at home in almost every aspect of it. (1979a)

Needless to say, intuition develops through the player's *activity* and pupils' mathematical intuition will only develop via the same route.

well as the teacher does. Why is this? Because the teacher through experience has developed an intuitive understanding of the subject — as the pupils will in time — if they are given time!

Pupils struggling with concepts which are too abstract and difficult will not develop intuition — though they may well develop a repertoire of tricks and pseudo-concepts as they attempt to 'get the right answers' to questions they don't really understand.

On the other hand, when *working* in situations and on problems that are within their reach, they naturally develop intuition and will be able to ask questions about them, think creatively and in that sense will be more highly motivated.

Good intuition is essential in *posing* problems. Intuition points to puzzling features of a situation and highlights features that the problem poser does *not* understand and it also suggests problems that will be interesting, intriguing and promising — so it is also related to aesthetic judgement.

The more abstract a topic, the harder it is to develop intuition, even for very strong students. It is much harder for most pupils to develop an intuitive feeling for algebra because it is so abstract, and so posing problems in algebra and investigating them is harder also.

Given good intuition, pupils will be better able to think confidently about a topic and pose their own problems — even if they cannot answer them all!

Conversely, without intuition the pupils is always struggling: he or she 'just' understands, can 'just' solve the exercise or the exam question if they can remember how it goes. But they can never be really confident, which is why lack of intuition is demotivating. A report of 1919 by the Mathematical Association argued that,

> We must prevent the suppression in modern youth of that naive intuition which according to Professor Klein was 'especially active during the period of the genesis of the differential and integral calculus'.

Why might teachers overestimate their ability to convey ideas in words? Partly, no doubt because of the widespread idea that understanding is merely procedural — this is what you do — just like operating a washing machine — select the correct program (algorithm) and so on ...

The proof that this is not sufficient is the very shallow levels of understanding that is the most that so many pupils acquire, after years of schooling. The obvious observation that pupils almost always end up knowing less than their teachers, because the teacher cannot convey in so many words his or her intuitive understanding, bears repetition.

If intuition could be conveyed in so many words, then you or I might be able to grasp the profound intuitions of great mathematicians like Atiyah or Arnol'd! Atiyah for one would disagree:

> It is hard to communicate understanding because that is something you get by living with a problem for a long time. You study it, perhaps for

years, you get the feel of it and it is in your bones. You can't convey that to anybody else. Having studied the problem for five years you may be able to present it in such a way that it would take somebody else less time to get to that point than it took you. But if they haven't struggled with the problem and seen all the pitfalls, then they haven't really understood it.

(Atiyah 1984:17)

Pupils in school have very little opportunity to become immersed in anything, starting with the difficulty that lessons last for anything from 30–35 to 45 minutes to an hour or so, and then stop. Shamshad Ersan (the billiards problem, page 142) did have a chance to become immersed in his problem. A small group of pupils spent about eight hours in total, only occasionally disturbed by visits from me to see how they were getting on. All the group kept going for this length of time, with no difficulty, without being as startlingly successful as Shamshad Ersan (1986a:14).

NOTE: Fishbein's (1987) intuitions are different from mine — his are *certain*: 'At every level of mathematical reasoning three basic aspects have to be considered: the formal, the algorithmic and the intuitive' and 'The intuitive aspect refers to the degree of subjective acceptance or rejection of mathematical concepts or statements.' According to this view, 'the intuitive component is expressed in intrinsic beliefs attached to these concepts and operations' (Fishbein 1987).

4

Appreciating = Knowing about Mathematics

In the teaching of mathematics it is possible to distinguish between three elements — facts and skills, conceptual structures, and general strategies and appreciation.

(Cockcroft Report:240)

As taught in the past (mathematics) has been informed too little by general ideas, and instead of giving broad views has concentrated too much upon the kinds of methods and problems that have been sometimes stigmatised as 'low cunning' ... It is sometimes logical, but the type and 'rigour' of the logic have not been adjusted to the natural growth of young minds ... *We believe that school Mathematics will be put on a sound footing only when teachers agree that it should be taught as Art and Music and Physical Science should be taught, because it is one of the main lines which the creative spirit of man has followed in its development.*

(Spens Report 1938:177; italics in the original)

Normal school teaching nods at conceptual structures but ends up by teaching facts and skills (poorly), largely ignores 'general strategies' and totally ignores appreciation. A grave mistake.

A basic principle of all teaching is to put the subject in context: this does not mean making it so directly relevant to the pupils' lives that they are prevented from understanding anything 'beyond their ken', as some right-wing critics of education would claim. It does mean making lots of meaningful connections.

An example: one chapter of our history textbook when I was in the lower school of Bristol Grammar School in the late 1940s was all about the brave and fearless ancient Greeks. I can still recall two of the pictures: the daredevil Alcibiades lying in the street, challenging the driver of a

Appreciation versus knowing how

'In the visual and dramatic arts, literature, in all sports and games, a clear distinction has always been recognised between the skill of the practitioner and the skill of the appreciative spectator ...

Of course, the amateur will benefit if he (sic) plays football, or tennis, or is a Sunday painter or has tried to write a novel or has studied science, but this gain is limited. No amount of dabbling at a low level by itself will give the amateur the appreciation of the professional's work which comes from a combination of his own limited experience and studying the professional in action — if necessary with a commentary to point up the salient features of the action ...

As a result, the number of people who appreciate a given field is far greater than the number who practice it in any professional sense ...

To treat everyone as if they will or should become very skilful themselves is not reasonable, not to say unfair. Rather, everyone should be given the opportunity to advance in skill and appreciation in parallel, especially initially when their level of potential skill is not obvious ...

In ignoring appreciation educators not only place the greater burden on the weakest shoulders; (naturally it is the strongest pupils who most easily digest a diet of nothing but facts and techniques); they also successfully hide the nature of these subjects as *activities* which are practised by *actual live human beings* ...

A note of warning. Let educators loose in 'appreciation' and before you can say *bureaucrat* it will have become another set of subjects. Educators will hunt around for a definition of each subject, (and a definition of appreciation itself, what it 'is') and having discovered what mathematics *is* or should be ... they will hasten to pass this information on to students as *facts*, in which they can be *examined*.'

(1977:10–12)

cart to run over him; and a Persian spy watching the flower of the Greek army exercising. The caption explained that the Persian little realised that these men would soon be defeating his mighty force.

This connection between ancient Greece and a small boy in mid-twentieth century Bristol is obvious: small boys — then as today — are likely to be into fantasies of heroism, bravery and power.

What useful and long-lasting connections might a maths course make? It is often supposed that a mathematical education enables you to, well, do mathematics, which suggests in turn that it is of little value to anyone who doesn't actually need to use algebra in their work, use geometrical ideas (because they have become a surveyor or architect) and so on. And even then it might seem they need only a very little specialised knowledge: it is not true that either surveyors or architects need to know many of Euclid's theorems.

Not so! What mathematicians carry with them into everyday life is more a large number of concepts — *general concepts* — that can be used to understand everyday problems and situations.

The football analogy

A high level of appreciation with limited technical skill is very common — which is one reason why football is a world-wide phenomenon. The typical fan's technical skill has rusted since his or her schooldays and was maybe never there if he or she disliked football but appreciation, if he or she goes to matches, will have increased.

In music, technical proficiency and appreciation are usually completely separated — most people play no instrument but they appreciate (certain types of) music.

To aid appreciation, the critic or commentator — including the authors of articles in the *New Scientist* or the *Scientific American* — mediates between the professionals and the public, while professionals themselves have their coaches.

How can the typical reader understand an article in *Scientific American* magazine? Certainly not because they have a detailed understanding of the subject matter: rather because they have *a general appreciation* of the theme.

Fortunately, also, there is much crossover between the appreciation of one sport and another: realise that footwork is an essential skill in tennis, and you at once generalise to other sports, and by analogy to finger work in playing a musical instrument. A little appreciation goes a long way.

Unfortunately, school subjects outside the arts curriculum are marked by an absence of appreciation.

The immediate visibility of an activity does not mean that all its features will immediately be equally visible. Not at all, some will only become visible as appreciation develops. My appreciation of football being strictly limited, when I hear a commentator talking about a match, or listen to the commentary on TV, I often fail to understand what the commentator means, while plenty of pupils will understand better then I.

The skill of a tennis player who strikes the ball right into the far corner is very visible, the skill with which he hits a backhand volley less so. Non-arts subjects are relatively invisible and mathematics is plausibly the most invisible of all (1979c).

Metaphors and analogies

Appreciation naturally involves metaphors because it is about making connections and spotting similarities in difference (and difference in similarity). The metaphoric eye is very good at seeing the general and shared features and is therefore very helpful to appreciation.

Metaphors and analogies are especially crucial in making links between the invisible aspects of mathematics and other phenomena and enabling teachers and their pupils to talk *about* mathematics.

In the claim, 'a triangle has many centres' the word centre is a vivid metaphor rather than a precise term. Anything which has some sort of 'centeredness' about it might qualify — including the circumcentre which for obtuse triangles lies outside the triangle! This metaphor of the 'centre' is at one and the same time subtle and profound, vague and imprecise and true without being a mathematical theorem.

The metaphor of the detective is an example of a familiar activity that simulates the *process* of doing mathematics. The mathematician solves problems like a detective, making observations, putting clues together,

The metaphor of a code

LJABO   UO   BPA   LJABO   UO   VPPW

LJABO   UO   TBZ   U   NP   AP   OVBPPW

EDPSWILO   LJGI   LI   DJVG   LT   SDJUM

SQA   LA   SDJUM   VJM   AJGI   ABI   OADJUM

U   JWTJZO   YUMC   ABI   BUCCIM   JMOTID

U   MIKID   OJZ   PB   MP   U   VJMA   OUD

(Based on 1979c:13–14; 1980–1983:#2, 9)

drawing false inferences — realising that he or she has done so — going back to the basics of the situation and so on.

There are other benefits from the detective and clues metaphor. For example it becomes natural to see that the fewer clues you have, the harder it is to solve the problem, suggesting the question, 'What is the smallest number of clues from which the conclusion can be deduced?' — without ever using the abstract language of axioms.

This is a variant of a code that I have often used with pupils who have never done problem solving before, that is, who are accustomed to learning merely from a textbook and from teacher explanation.

The purpose of the code as I explain it is to get them thinking like mathematicians, searching for clues, putting the clues together to make sense, drawing rational conclusions, guessing, possibly discovering that their guesses though plausible are mistaken and persevering even when they seem to be stuck.

The self-motivating effect of such codes is that they close in on themselves, as it were, like crosswords. The feeling of achievement that you get when you are approaching your goal is strong and visible.

Anxiety of *understanding about*

Anxiety may be caused by failure, or anticipation of failure to master contents but it can also be caused by a failure of understanding *about* the subject. Pupils who move from (for example) a chapter on fractions which is genuinely difficult to a chapter on sets which they ought to find cognitively very easy (albeit pretty pointless at their level, as we have already discussed) may find even sets difficult if that is their *expectation*.

Children need overviews which themselves can be from different points of view. A topic might be put in context of the physical sciences, an artistic context, it might be given a practical context, a vocational background, a social perspective, viewed in the light of another mathematical topic or a mathematical idea, such as an important general concept.

The businessman analogy

A businessman preparing a course for trainees ensures clearly defined objectives, a clear sense of intrinsic achievement to look forward to and defines the end of the course so that trainees can appreciate its approach. The elementary psychology involved is transparent.

Businessmen can also take for granted a high level of appreciation of business in very general terms, without which trainees would be completely lost.

Schools generally follow none of these principles. In addition, the background to most subjects is not so familiar to pupils. Many school children are indeed extremely lost.

The real world, the world of the child, is a closely woven web of meanings and relationships and connections, without which it would be literally meaningless.·

When children go to school, most of these connections are broken. The world of academic subjects is very low on connections, therefore very difficult for the child to grasp.

These many and varied connections constitute the effective interpretation of 'relevance'. Narrowly interpreted, calls for relevance rightly draw the fire of conservatives. But interpreted as the general connectedness of ideas and experience, it embraces all those relationships to tradition, to culture in general, which are beloved of traditionalists.

This web of connections inevitable involves affective connections, as well as conceptual, if that separation can sensibly be made, which is improbable. Emotional meanings are as important as cognitive but are largely missing from academic school studies (1977:13).

Learner drivers

Learning to drive a car is a common real-life learning activity. All learners know a great deal about cars before they even start to learn, having been driven often and having read about cars and watched them (examples of *surreptitious* learning).

They also have other experiences which will give them an idea of what happens in particular circumstances, for example when cornering, when the car and the driver are swung to one side as if on a fairground roundabout.

Car driving is also a desirable achievement. Just as important, there is very little pressure to learn and it is widely accepted that some people take far, far longer than others. No one likes taking their test 20 times and a teenager might be embarrassed to do so but everyone else can laugh off their slow progress.

Finally, *the examination is taken when the pupil is ready*, not when the teacher decides.

Spot the difference

- School courses have no well-defined objectives, even for most teachers, let alone for the children. I do not suppose that examinations in the fifth year is any sort of objective for at least the first three years. Many children seem to suffer from not knowing where they are going or not seeing why they are doing what they are doing. I sympathise with them.
- The school course has a start and finish but for all practical purposes these are so far separated as to be meaningless.
- This is not compensated for by any definite, recognisable progress within the course. At no stage can the child feel, or know, that he has reached a certain level of achievement.

What would a mathematician do?

A pupil had asked a question about a problem on which they had got stuck, to which I could have answered (for example): 'Well, how about making a table to organise all that information? Then you can see what you are doing.' Instead, I replied on the spur of the moment, 'A professional mathematician would put all the information into a table, so that he could see it better.'

This was a natural response given my interest in 'putting pupils in the picture' and hence developing their *appreciation* but I had never thought before of putting it so bluntly and directly. I got into the habit of prefacing comments, 'A professional would … ' and was soon rewarded when a pupil announced, 'Mr Wells wants us all to be mathematicians!' (1994c).

This developed into a *spiel* to introduce the theme of problem solving to pupils which incorporated an angle on 'What a professional mathematician would do' and summarised a few basic points. I would then re-emphasise these points and add others when necessary: many general concepts represent the 'proverbial wisdom' of any experienced mathematician, professional or amateur and opportunities to introduce them occur frequently.

The spiel went something like this:

- If you are given a choice of problems and the one you choose turns out to be too difficult, or you simply dislike it, you give up and choose another.
- Mathematicians don't bang their heads against a wall trying to solve problems that are too hard for them, and they don't waste their time on problems that are too easy to be challenging.
- Pure mathematicians have a bigger choice and more TIME: applied mathematicians are working in a particular field, and may be required to produce 'answers' (= solutions) under stricter conditions.
- Applied mathematicians who work in industry have to tackle the problems that are a part of their work, but mathematicians who are not paid by someone else have a more-or-less free choice.
- It is impossible to draw a dividing line between pure and applied mathematicians.
- Finally: if I give you a problem and you can't solve it, that that is my fault, because my judgement of what you could do was wrong.

Levels and perspectives

Pupils themselves need to understand the idea that there are different levels of understanding of a topic and that they are being taught with a particular purpose in mind. Pupils need the level of understanding that is appropriate to the situation they face.

Fortunately, pupils find it very easy to understand the difference between just relying on a rule which gets the right answer though they don't understand why; understanding why, step by step (usually with limited confidence), and finally, understand so well that everything is clear and obvious — even if you have difficulty in explaining to someone else WHY it is clear and obvious, which is another problem!

One response to the invisibility of mathematics is to teach pupils what they cannot see for themselves and will never find in standard textbooks. We can teach them ABOUT mathematics, ABOUT how it is used, ABOUT connections between topics, ABOUT thinking.

We can teach them to appreciate mathematics and its contents, in addition to helping them to learn the contents and processes themselves.

There are many ideas that can be shown to pupils, for example, that there are many ways to solve any basic arithmetical problem. This is an idea that, unusually, is indeed taught to most pupils. A much more advanced example is that 'an infinite series of rational fractions can have an irrational sum'.

General concepts

Traditionally almost every subject, including mathematics, has been taught as a collection of facts, or techniques.

But our knowledge of the world does not consist largely of facts. It is not certain and precise but more or less vague, uncertain and probabilistic. We are guided by hunches, feelings, judgements and guesses.

Heuristics, exceedingly general in form and wide in scope, mostly apply to problem solving in general and to other subjects apart from mathematics and so are only one aspect of non-factual knowledge.

Most of our mathematical knowledge is neither as widely applicable as heuristics, nor as narrow as the facts in the textbook.

Rather it consists of *general concepts*. These ideas do relate to content. Yet they are not mathematical facts in the way that theorems are. For example, they are not proved.

Generally speaking they are vague, imprecise, require interpretation, are potentially useful as well as misleading (especially if taken too literally) and can often not be described as either true or false without qualification.

Nevertheless they do correspond to an important aspect of the mathematician's or mathematics teacher's knowledge, which the teacher needs to pass on as far as possible to the students.

General concepts are a large part of the everyday currency of mathematical thought, talk and discussion, whether we are aware of them or not. By being more aware of what we are doing and how we are doing it, we as teachers can teach, and help our pupils to learn, more effectively (1987d; 1988g).

General concepts have other important characteristics:

- They can often be thought of as a wit and wisdom, the proverbial wisdom of mathematicians.
- Proverbs often employ figures of speech — e.g. the use of 'centre' to describe a variety of points of a triangle!
- They are also a wonderful stimulus to problem posing, just because their combination of truth with vagueness demands interpretation: what does 'centre' of a triangle *mean*?
- They are often not statements of fact, but rather fallible guides and pointers based on wide experience.

They represent, in words, some of the feelings, intuitions, expectations and seasoned judgement of the mathematician. They are specific and dependent on the experience of many mathematicians, so pupils cannot be expected to infer most of them from their own limited experience. They are the kind of wisdom that a tennis coach bring to a tennis player.

Such ideas *about* mathematics help to put topics in a context and so help to answer the question 'What's the point, sir?' and are very strongly motivating.

Guiding and coaching pupils

Such general concepts are especially important for the weaker pupil. There is an analogy here with the question as to whether students benefit from studying Polya's solving-problem heuristics as presented in *How to Solve It* and his other books. It has been claimed that using Polya's heuristics with young mathematicians does not help them. Well, no surprise, heuristics are most important for pupils who are not talented, do not find maths easy and so will not 'work them out for themselves'. With typical pupils, in other words, who can be helped to explore their problems with helpful heuristics.

Most students are not that talented and need more help: they need that help as or even before they start to tackle a particular problem — not a solution, they don't need to be told *How to Solve It* — but they may well need to be 'put into the picture'.

Talented pupils naturally tend to select problems and use methods that fit their natural way of thinking. Weaker pupils with less confidence and a greater tendency to rely on the teacher will not do so.

Here are some general concepts about mathematics itself. We have commented on some of them already. They are followed by General Concepts about numbers, problems, and methods and finally algebra contrasted with geometry.

General concepts about mathematics

Mathematics is difficult.

Mathematics is mostly invisible.

You can often know what you can't write down.

Many mathematical situations and ideas can be presented either in a dynamic active form or in a passive and static form.

The best mathematics is usually simple.

'Simplicity' is often however, a moot concept. Is it simpler to multiply by eight by doubling three times — which you might be able to do in your head or by writing down each doubled number from the front — or to do an ordinary sum using your 8 times table?

In many textbooks, changing $(x+1)(x+2)$ into $x^2 + 3x +2$, for example, is called simplification, or simplifying the expression with brackets. Is 'simplification' a good word to use? In what way is the last expression simpler than the first (*The Problem Solver*:#6)?

Even the purest mathematics tends to become useful sooner or later.

Giant numbers with only two prime factors are used in cryptography. If you could find a method of finding the factors of very large numbers, you could break important codes and become famous and a millionaire.

Mathematicians do not agree on what is good mathematics.

Some professional mathematicians think that Euclidean geometry is too simple and old-fashioned. Others think that it is still beautiful and illustrates many important mathematical ideas.

Mathematics can be very simple indeed — and simpler than you expect.

We have quoted Paul Halmos on general set theory being pretty 'trivial stuff really'. Halmos was telling readers of his book *about* its subject in order to help them to understand it as easily as possible. Had he not done so they might have found this easy topic harder than necessary.

Metaphors and analogies in mathematics can be exact.

Mathematical objects can be looked at in different ways.

Proof and certainty are possible in mathematics.

Mathematical language can be interpreted in different ways.

Funny things happen at infinity. Infinity is weird.

0 and 1 are special (and especially confusing).

General concepts: numbers

Patterns in numbers are everywhere.

Odd and even are important everywhere.

Small numbers are very important.

They can also be attractive, as are very large numbers, both being very good examples of 'extremes'.

Fractions are ambiguous.

There are more numbers than you think.

You can only write down small integers.

Whatever number you write down, you can write down a larger one.

And whatever number you write down less than ten, you can write down a larger number which is still less than ten. These puzzling ideas can be thought of as riddles, and have much the same attraction.

You can't write down the smallest fractions or decimals.

Whatever number you write down, you can write down a smaller one.

Whatever number you write down, you can write down a number that is very slightly larger or smaller.

The number 0 appears often in mathematics.

General concepts: problems

Problems can be divided into types: some which seem to be different types, are actually the same type.

Many problems cannot be solved and you can prove this.

Interesting things happen at extremes.

Kepler noticed that at maxima and minima the function changes more slowly.

Even when everything seems to change, something stays the same.

A hard problem can look very much like an easy problem. You don't discover the difference until you try to solve it.

What is the square root of 169? What is the square root of 170? What is the square root of −169?

Some problems can be solved by algorithms, others cannot.

Problems are often ambiguous: you have to decide what they mean or what you will MAKE them mean.

'What is the centre of a triangle?' Here 'centre' is ambiguous: you can find more than one centre for a triangle.

If a problem has several solutions, they are connected.

'Find a circle which touches all three sides of a triangle.' There are four such circles and their centres and other features have many properties.

General concepts: methods

If you know a pattern, you can make predictions.

If you just happen to know the squares of the integers not just up to 10 or 12 but up to, say, 25, then you can predict the answers to products such as 17 × 13 or 19 × 23 long before someone who is 'doing the sum' in their head gets the answer.

Every new idea creates a host of problems.

Idiosyncratic methods can be very illuminating.

If you only want an approximate answer, step by step, you can often afford some small mistakes on the way.

Linear usually means simple.

Mathematics is beautiful. You can expect to find patterns everywhere.

Put a pattern in, get a pattern out.

Everything is equivalent to something else.

Some problems can be solved by algorithms: you follow the algorithm exactly and do exactly as you're told — and out comes the answer, or answers.

Some problems need techniques, such as algebra, or trigonometry. You are not told exactly what to do, you have to decide what moves to make yourself.

Hard problems can only be solved by the mathematician's creativity, by thinking of new ideas.

It is worth looking for analogies of features in two dimensions, in one and three dimensions.

You need to choose the 'appropriate' method for each problem, and this method depends on the problem.

First because if you choose the 'right' method you solve the problem relatively easily, but choose the wrong method and you get into a mess — even if the solution eventually appears; second, because the 'right' method will inevitably reveal the structure behind the surface of the problem.

Any number or function can be represented by an infinite series.

Maxima and minima don't always need calculus.

Some methods of solution are faster than others; some are more general.

Symmetry is usually helpful. Look out for symmetry.

There may be more than one structure underlying the same situation.

There may be more than one way of looking at a situation.

Many problems can be solved by many methods

The puzzle of finding the limit point of this spiral illustrates both of the last two general concepts: the sequence can be summed in several ways, by using similar triangles, by calculation or by noticing that one part of the figure is similar to another.

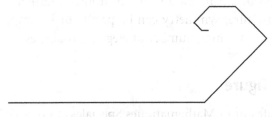

Where does this spiral end?

(1980–83;8, 25–26; 1982:12)

Many mathematical problems can be represented by diagrams.

Many algebraic problems or situations can be represented by diagrams.

Many proofs can be put into a diagram or figure.

Many problems in geometry can also be interpreted as problems in arithmetic or algebra.

General concepts: algebra and-versus geometry

Geometry is the least abstract form of mathematics: this means that it has direct applicability to everyday life and also that it can be understood with less intellectual effort.

(Atiyah 1982)

Geometrical diagrams mean a lot.

Geometry emphasises meaning and algebra emphasises calculation.

Geometrical symmetry can be incomplete (the Vecten figure, below).

You can manipulate algebraic expressions even when you have little idea what they mean.

You need imagination in algebra to think of the next best move — unless it is just a matter of technique.

Geometry includes an astonishing variety of figures and objects.

Proofs in geometry may consist of just seeing a figure 'in the right way'.

The Vecten figures (below) illustrate several general concepts: you need to choose an 'appropriate' method for each problem; tessellations highlight pattern and structure; symmetry can be partial or incomplete and some figures have an astonishing number of elegant properties.

The Vecten figure

Vecten was professor of Mathématiques Speciales at Lycée de Nîmes when in 1817 he wrote the paper that introduced this figure which has become very popular:

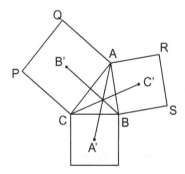

The original Vecten figure

Many of the properties of this figure can be proved by examining this tessellation in which the Vecten figures is embedded (2008b). The tessellation has rather a strong symmetry though this is limited because the original triangle is not an equilateral triangle. The symmetry would be stronger if it were equilateral.

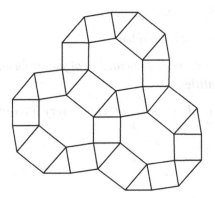

The Vecten figure, tessellated

The next figure shows some different embedded properties of the Vecten figure:

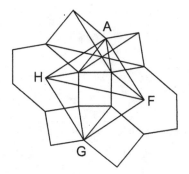

The Vecten figure: proof from tessellation

More general concepts

Algebraic proofs can be totally unilluminating — they simply show that the conclusion is true without explain why.

Algebraic expressions are in a code.

Algebraic manipulation is much like playing a game: there are certain moves that the rules allow you to make and others that are not allowed.

You may have to decide from the context whether a letter stands for a constant or a variable.

Algebra has fewer objects than geometry.

Symmetry in algebra is often hidden.

Many geometrical theorems if presented in algebraic form, turn out to be not very interesting identities.

Geometrical proofs of algebraic results can be very illuminating.

5

Motivating Emotions: Beauty and Aesthetics

We need not hesitate to put the pleasure motive among the foremost ...
It would be surprising if mathematical ideas, properly presented, fail to
give pleasure to the average student ... A mode of thought so fundamental might be expected to arouse a sympathetic vibration in the intellect.

> (*The Teaching of Mathematics in
> Public and Secondary Schools* 1919)

Mathematicians have frequently lauded the beauty of mathematics and
their best known quotes have become clichéd, so here are just two of the
most famous:

> I think it is correct to say that (the mathematician's) criteria of selection,
> and also those of success, are mainly aesthetical.
>
> (John von Neumann 1947)
>
> It is true aesthetic feeling which all mathematicians recognise. The useful
> combinations are precisely the most beautiful.
>
> (Poincaré 1930/2003:59)

Poincaré also gave a warning which will be recognised by anyone who has enjoyed problem solving:

> When a sudden illumination invades the mathematician's mind ... it sometimes happens ... that it will not stand the test of verification ... it is to be observed that almost always this false idea, if it had been correct, would have flattered our natural instincts for mathematical elegance. (*Ibid.*)

Pure mathematicians especially have enthused over the extraordinary beauty of mathematics — and sometimes been led astray as Poincaré warned — but the use of mathematics in the physical sciences has also astonished and delighted, the more so as the sciences generally have become more and more mathematical.

Here is V.E. Johnson on *The Uses and Triumphs of Mathematics* (1889):

> That department of Mathematics known as the conic sections, has an especial beauty, from the fact that those curves with which it is concerned are the curves in which the planets, comets, etc., move round the sun ... but as the principal use of the science of Mathematics lies in its applications to the Sciences and the Arts, so there, I have said, also lies its chief beauty.

Henry van Etten in his *Mathematical Recreations*, published in London in translation in 1633, also emphasised applications and the sciences. His title page listed these branches of the Mathematics, illustrating why it was a plural proper noun:

> Arithmeticke, Geometry, Cosmographie, Horalographie, Astronomie, Navigation, Musicke, Opticks, Architecture, Staticke, Mechanicks, Chimestrie, Waterworkes, Fireworkes.

How fascinating! How exciting! What vistas of understanding, power and achievement he opens up to the beginning student! Van Etten's attitude as compiler and poser of his problems is sketched in his 'By way of advertisement' in which he says that he has not explained the solutions not least because, (these things) ought to be concealed as much as they may, in the subtill-tie of the way: for that which doth ravish the spirits is, an admirable effect whose cause is unknown: which if it were discovered, halfe the pleasure is lost.

But, we may suppose today, the other half retained and new pleasures added. Michel de Montaigne (1533–1592) recorded that,

> Jacques Peletier was telling me at my house that he has found two lines approaching each other, which, however, he established could never succeed in meeting except at infinity.
>
> (Lionnais, Le 2004:v.2)

Surprising indeed! School pupils who have only graphed straight lines and parabolas are easily impressed when they first meet a curve with asymptotes: instead of heading steadily towards the edge of the graph paper the asymptote races off the page! An asymptote also has the strange feature that the curve gets closer and closer to it, but never meets it — an example of *limits*, another counterintuitive topic which easily intrigues children.

This double effect of expectations met and expectations surprised is especially strong in mathematics just because our confidence in maths is so great, so our expectations are perfectly met — except when they are amazingly confounded.

This creates a potential double bind for teachers: their pupils need experience *before* they can develop expectations and then have them confirmed or confounded: so how can teachers give pupils beautiful experiences to motivate them from the start? An answer is through the use of mathematical recreations that are already more or less 'familiar'.

Jacques Ozanam wrote in his *Cursus Mathematicus* (1712) that:

> The greatest part of the Lovers of Mathematics are won to that Science by its sensible Beauties only, they are taken by the Wonders that it will works and delighted by its admirable *Phaenomena*; they are willing to know what they have admir'd, to perform those Things that at first they could not account for; and take pleasure in surprising others, as themselves have been surprised.

Paul Carus (Andrews 1908) explained the fascination of magic squares:

> The peculiar interest of magic squares … lies in the fact that they possess the charm of mystery. They appear to betray some hidden intelligence which by a preconceived plan produces the impression of intelligent design, a phenomenon which finds its close analogue in nature.

There is a natural pay off between the pleasures of mystery and wonder, and the satisfaction of understanding, as Edward Titchmarsh illustrates:

> Here, one felt, was mathematics really happening before one's eyes. The theorems were obviously of the highest interest, and yet it was a book in which one did not have to rack one's brains to discover what the author could possibly mean. The older English mathematical books were full of mystery and wonder. With Professor Watson we reach the period when the mystery is dispelled, though the wonder remains.

(Titchmarsh, in Coulson 1969:234)

Most pupil textbooks, unfortunately, convey none of this excitement, offer no sense of mystery, present no enticing prospects to the pupil, and as we have already noted, completely omit any reference at all to the sciences as if mathematics were an isolated phenomenon, unconnected with the real world (other than via everyday arithmetic) or with real life and by implication unconnected to the lives of our pupils, in which case the more feisty among them will be repeatedly tempted to ask, 'What's the point?'

Pupils should be encouraged to use their own aesthetic judgement, even if it is not the teacher's. In particular, if they are given choices of problems to tackle, which they try to solve as far as possible by their own understanding, then not only are they likely to be more motivated by the problem that they chose themselves but they will develop their appreciation of different kinds of problems and methods of solution and they will naturally develop their own judgement of problems and their own aesthetic feeling.

Once again, science teachers have a natural advantage: there are so many beautiful, amazing and fascinating objects and phenomena in the natural world which they can exploit to motivate their pupils. The most

Pupils and aesthetic appreciation

'Do we want our pupils to appreciate these (aesthetic) features of mathematics? If so, then they must meet them frequently, and in relation to every aspect of their mathematical activity.

Do the problems that they tackle grab their interest? Is Mary delving into the periods of decimals because she is intrigued by them, or because her teacher thinks she ought to?

Is John spending ages over a problem because he refuses to let go of something that he is still making sense of?

Have the class been introduced to a variety of mathematical objects and images? Have they created many of them themselves? Do they realise that when they enjoy these objects and images, they may well be sharing their enjoyment with millions of people throughout the centuries?

Were they recently amazed by a mathematical object they had never seen before?

Do they consciously try to make their solutions as clear, and concise, and elegant as possible?

Do they share their work with others in the class? Are they aware that there is a wealth of intriguing, attractive mathematics that has been discovered and recorded by past mathematicians?'

(1989b:35)

remarkable phenomenon of science are well-known, sometimes because like rainbows, snow storms, colds and fevers, vast bridges and crazy towers, we meet them in everyday life and sometimes because they appear in the media.

Fortunately, mathematics teachers can exploit some simple scientific phenomenon to motivate their pupils, as we shall illustrate in Ch. 9.

Beauty in school mathematics

... a campaign to promote *Multilink* as an all-purpose learning aid suitable for a wide range of topics and ages. It now comes in two colour schemes: black, white and grey for secondary schools, and this familiar, bright, multicoloured version for primary schools.

(Review of *Multilink*, *TES* 19 May 1989)

In the early twentieth century, the first of John Perry's 'obvious forms of usefulness' of mathematics was, 'In producing the higher emotions and giving mental pleasure. Hitherto neglected in teaching almost all boys.' (1986a:21)

We might agree with Perry that beauty ought to feature in school mathematics but does not. As André Mack puts it, discussing aesthetic aspects of mathematics:

> work in this area has proven to be of little use to the mathematics educa-
> tion community beyond a cursory acknowledgement of the existence of
> a specific mathematical aesthetics … The pervasiveness of aesthetic
> perception in the practice of mathematics makes its limited role in math-
> ematics education somewhat of an unfortunate anomaly.
>
> (Mack n.d.)

One difficulty is that the language of 'beauty' and 'elegance' does not come naturally to maths pupils. H.E. Huntley in his book *The Divine Proportion* describes how his sixth form pupils once laughed at him when he referred to Pythagoras's theorem as beautiful. (Perhaps he was too blunt or took them by surprise.)

Another, more severe, difficulty is that mathematics in schools, at least in secondary schools, consists largely of utilitarian instruction by the teacher to enable pupils to 'do' an exercise correctly, followed by the pupils demonstrating that they do just that, whereas the beauty or 'coolness' of mathematics appears when adults and pupils are tackling mathematical *ideas*, solving puzzles and problems and being *active*.

G.H. Hardy made a connection between the puzzle columns in the popular press and mathematics:

> Nearly all their immense popularity is a tribute to the drawing power of
> rudimentary mathematics, and the better makers of puzzles … use very
> little else. They know their business; what the public wants is a little intel-
> lectual 'kick', and nothing else has quite the kick of mathematics.
>
> (Hardy 1941/1969:86–88)

This suggests another link, to psychologists who have studied the Aha! experience: the penny drops and you finally 'see it', whatever 'it' may be and you get a 'buzz', plausibly much like the 'kick' that Hardy was talking about. This response depends on making connections, spotting surprising relationships and being creative and is extremely rewarding.

Pupils, mathematics and beauty

Today's pupils are capable of discussing the beauty of mathematics if it is appropriately introduced though (if unprompted) they may well describe an image, argument or conclusion as 'neat' or even 'cool'. Such responses seem indubitably aesthetic, though the language used is not traditional. The following questions and pupil responses about 'beauty' in mathematics were from conversations between the author and some sixth form students the West Sussex Institute of Higher Education:

Question: 'A curve can look pretty or beautiful on the screen. Can an equation or something which is hard to understand and which you cannot see with your eyes, also be elegant or beautiful?'

Wei-Wah Tang: 'Yes. This is a lot to do with something like literature. You can say, 'That is a beautiful sentence', the way it is structured, the way it is built up, the way it defines something, describes something so perfectly, and so on — everything seems right about it, it seems beautiful or elegant. First you see beauty as what you can see with your eyes, but then as you read and develop, your perception expands, its boundaries expand, and you can enclose other things as beautiful.'

Daniel Smith: 'You mean, like, you're really excited about solving a problem and went out and told all your friends you have just done a wicked problem.'

Alison Hird: 'It's like … sometimes you look at all these wonderful graphs you get, and you think, 'Aren't they really lovely!' And everything like that, they are fascinating … It depends which bits of mathematics you are talking about, because some of them can be tedious to go through (but) the more visual stuff can be quite beautiful.'

(Question): 'Would you think that Einstein's '$e = mc^2$' is beautiful? Or not? Or just amazing?'

Tracey Martin: 'I'd say that it is amazing, rather than beautiful … I admire him for what he did, and I would say he was amazing or brilliant.'

Daniel Smith: 'I would say it depends on whether you are interested in the subject and can really get inside it, and think about it deeply, really get involved with it … If you can then … you would use the same language that you use for something else, but, it is much harder to get inside, say, the formula for something and see the elegance of it … than it is to look at a painting and think, 'That looks good!' Once you have got to a high enough level, you will find it much easier to look at something and think, 'That's good'.'

Robert Geleit: 'In mathematics I've never come across a situation like that, but in computers I have. Many a time, I have looked at a program that I've just listed up, and it is so well structured, you know, it's just, amazing. So far, I've found computers more beautiful, but then again, over the years I have been at school I have put much more time into computers than into mathematics.'

Alison Hird: 'Looking at a painting is very superficial, it is subjective rather than an objective (view), and if you get into maths it is far more objective, because you have to sit down and work through it methodically … The further you get into it, the more fascinating it can become, but a painting can be beautiful without really understanding the artist's meanings and emotions, while he's painting it, whereas, with a formula you have to go deep into it, you can't just look at the formula and solve the problem just like that, you have to actually work through, and that provides a far clearer view of beauty, if you like, than just looking at a painting and saying, 'That's lovely!'.'

(1989a/1990, v.3:33, 35–36)

Individual differences in appreciating 'beauty'

Beauty is in the eye of the beholder.

Like the teachers and mathematicians already featured, pupils differ in what they appreciate, as all teachers know. Some are fascinated by visual patterns but dislike and show no aptitude for algebra: others may be the reverse. As in every art and science, aesthetic appreciation varies between individuals though some characterisations are widely shared while others are more idiosyncratic. Here is Douglas Hofstadter on what makes a good problem:

> A good problem is one that mixes order and chaos in a deep and subtle way, and that fires the imagination for that reason ... In solving a good problem one discovers some wonderful and totally unexpected regularity when one expected nothing and on first sight saw only a jumble. The example of Morley's theorem in geometry is a good one.
>
> (Hofstadter p.c.)

To my mind — but my own reaction may be idiosyncratic! — this elegantly summarises much of the motivation of mathematics, though I would add that it is important that the mathematician finds this mixture of order and chaos as he or she explores some abstract miniature world, an activity that not everyone enjoys in the first place.

The physicist and mathematician Roger Penrose could have been responding to Hofstadter when he wrote of the standard square lattice:

> There is no doubt that it is something simple. But I do not really feel that there is much beauty about it. As a pattern it is just boring. It may just be a question of familiarity, of course ...

Penrose then turned to the famous non-repeating tilings of the plane which he admitted he much preferred but which he also discovered (Penrose 1974)!

Which is the most beautiful?

This quiz originally consisted of 16 propositions to be rated from zero to ten. It was presented to an audience of teachers at the Edinburgh 1992 Mathematics Teaching Conference and later enlarged to 24 items and published for professional mathematicians in the *Mathematical Intelligencer* with an invitation to rate each item and to comment on their own personal responses (1988e; 1990b).

Which is the most beautiful?

These are some of the problems presented to the audience of teachers:

J   There are an infinite number of prime numbers.
K   There are 5 regular polyhedra.
L $1 + 1/2^2 + 1/3^2 + 1/4^2 + 1/5^2 + 1/6^2 + \ldots = \pi^2/6$.
M   There is no fraction whose square is 2.
N   The four-colour theorem: every plane map can be coloured with four colours.
O   It is impossible to find three crossing-points on a square grid which are exactly the vertices of an equilateral triangle.
P   At any party, there is a pair of people who have the same number of friends present.

Here are the ratings for these seven items from seven members of the audience:

J	K	L	M	N	O	P
0	0	5	0	0	8	2
7	4	3	6	9	2	7
3	6	7	1	2	5	8
0	8	0	0	6	0	8
8	9	9	8	6	8	8
9	9	5	2	6	2	7
5	1	8	4	6	8	10

They are indeed varied and there were audible gasps at several points as the numbers were read out.

The results were instructive and consistent with the data in the box. One respondent gave every theorem a zero, claiming that 'Maths is a tool. Art has beauty', so his contribution was ignored as off the scale. Among the features mentioned by correspondents were simplicity and brevity, surprise, depth and the effect of changing the form of a theorem, as if the rating for beauty might be reduced for a different expression of the same claim. One respondent admitted that his ratings varied with his mood: he rated one theorem as a ten, but then changed his mind and rated it two! We can expect to find the same variability among teachers and pupils.

Components of mathematical beauty

We shall assume that like every other factor in the classroom situation, affect can be manipulated by the teacher, created, directed, played down, played up, in order to aid pupils' learning.

This suggests that if we turn from rather abstract concepts of 'beauty' which seems not to come naturally to pupils, to the components that are commonly supposed to explain or create it, the picture may change. If puzzle solvers get a kick from their puzzles and pupils sometimes find results 'neat' or 'cool' and can find them surprising or mysterious, then perhaps we should start from the pupils' actual reactions and not from philosophical conceptions of 'beauty' which mean nothing to them.

There are many such 'aesthetic factors' which can be created within school mathematics in forms that pupils can recognise. These features, expressed as nouns or adjectives, have been *roughly* grouped here by similarity — readers may prefer a very different classification.

We could be talking about an image or picture, an object, an idea, a problem, a claim, an argument, a phenomenon, a solution or a proof.

They are divided here into three groups, the first emphasising more affective responses such as surprise and astonishment, the second pattern, order and meaning, and the third focusing on power and effectiveness. A final, single, component is less an aspect of mathematical beauty than a response to it: the sense of challenge that we hope pupils will feel.

Surprise

curious	fascinating	wonderful	mysterious
intriguing	inexplicable	unexpected	unusual
amazing	astonishing	weird	strange
puzzling	impossible	paradoxical	exceptional

We are often challenged by the curious and fascinating, the surprising and the weird — we want to explore further, find out more, suck-it-and- see, test whether it really is true and we can get very strong emotional reactions.

Bertrand Russell writing on 'The Study of Mathematics' suggested that,

(The learner) should be instructed in the demonstration of theorems which are at once startling and easily verifiable by actual drawing.

(Russell 1907:32, 41)

Surprising, startling, intriguing and related reactions depend, however, on past experience. A result is startling against the background of the pupil's expectations which are a function of experience.

In Euclidean geometry that suggests results such as concurrency and collinearity which pupils can find surprising with almost no background understanding. The best-known examples are the various centres of the general triangle.

The figure on the left, related to the Vecten figure (page 84–85) shows two squares with a common corner: this property is enough to force all four lines to meet at 45°. Surprising?!

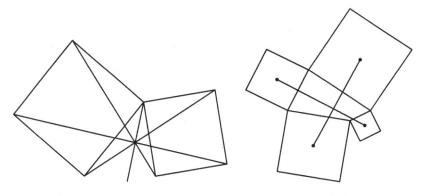

Two squares share a corner   Van Aubel's theorem

Van Aubel's theorem starts with an irregular quadrilateral — but construct squares on all sides and another pattern appears, slightly reminiscent of the Vecten figure: the two lines joining the pairs of opposite centres are perpendicular and of equal length.

Notice that pupils do not have to be able to prove such results, any more than science students must be expected to explain theoretically all the results they see in the science classroom or in amazing programs on TV. The positive results of surprising pupils with extraordinary facts and figures are much stronger than any disappointment that some pupils may feel because they cannot see why the result is true, and those pupils who feel most strongly that 'they want to know' are the ones most likely to develop the mathematical strength to eventually find out. We hope that pupils will often wonder why — but they know from their own experience that 'There are more things in heaven and earth ...' than they can possibly understand.

Opportunities for *simple* proofs for primary and secondary pupils are more likely to turn up in their own attempts at solving problems as we shall see in Ch. 6.

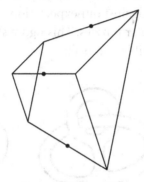

Similar triangles, same orientation

Sometimes the most familiar figures can spring surprises. The figure above shows two similar triangles with corresponding sides parallel. Of course the marked midpoints form a third similar triangle.

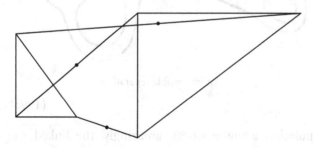

Similar triangles, different orientation

In this figure, the original triangles are similar but oriented 'at an angle' yet the midpoints (and therefore any other proportional dividing points) still form a third similar figure. (1988c:149. This feature is related to Napoleon's theorem.)

Simple calculations can also be very surprising: $\sqrt{2}-1$ and $\sqrt{2}+1$ are reciprocals. Odd? 142857 × 9999/999 = 1429857. A 9 has appeared in the middle. Coincidence?

The power of the impossible

We are easily puzzled by the apparently impossible, the paradoxical and the exceptional which go against our expectations. These emotions are the foundation of many popular puzzles. This figure shows one manipulation that appears to be impossible, but isn't:

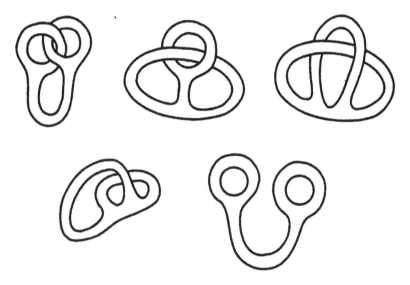

An impossible separation

(1995b:302–326)

The manipulation actually works: amazingly, the linked rings, if they are regarded as made of plasticine, can be separated by one continuous transformation. However, there are many other well-known puzzles that appear almost as baffling:

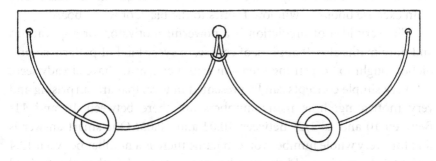

How can the rings be brought together?

This puzzle can be bought in many toy shops: the trick is to get both rings on the same side. It will not appear mathematical to most buyers but it is — it's a topological problem. Topology is full of surprises, even to professionals which is why there is a whole book, *Counterexamples in Topology* (Steen and Selbach 1970) suggesting that many hypotheses and plausible conjectures in topology are in fact false.

Presentation, surprising

Textbooks, ironically but typically, seem to go out of their way to *not* induce surprise in their readers, but teachers know that they can grab and hold the attention of any class by offering the surprising, even the amazing, not least by a suitable presentation. If Pythagoras's theorem is introduced as a dry statement about right-angled triangles to pupils who lack the experience to find it surprising, the presentation will fail and pupils' motivation will be minimised. If, however, Pythagoras's is used to *predict* the distance from one corner of the classroom to the opposite corner, pupils will start by being surprised and, all being well, conclude by understanding as well.

The idea that there is a *hidden connection* between the lengths of the sides of a rectangle and the length of its diagonal is highly motivating. (Starting with a rectangle is simpler, because the shape is much more common.)

Pupils can then, instead of going through a textbook exercise, practice their skill at prediction by using a variety of rectangles in the classroom, from exercise books to window-frames to the black or white board.

The very idea of prediction is a powerful motivator for applications and pure maths also. It applies, of course, to every kind of pattern, so prediction ought to be a strong theme in arithmetic, as we have already seen.

Very simple concepts can be presented in ways that are surprising and very motivating: how many numbers are there between 10 and 11? Between 10 and 10 1/2? Between 10.04 and 10.05? One smart answer is that for every whole number you can name there is a number between 12.4 and 12.5 — you just add the number at the end to lengthen the decimal part — but then some numbers when added, such as 1254 and 12540 will give 'identical' solution. Weaker pupils can find 'simpler' numbers to answer the question and still come to the same conclusion — there are as many as you want, or an 'infinite' number!

Order, pattern and structure

order	balance	symmetry	pattern
structure	invariance	connections	ambiguity
relationship	analogy	generality	unity-in-variety

Mathematics has been described as the study of patterns (which begs the question who creates the patterns and where they come from) and patterns are, of course, as old as man. Visual patterns appeal to everyone, tessellations especially, but they also raise many questions: they are curious and mysterious.

Visual patterns are especially 'obvious' but even the most simple may conceal more than they reveal at a first glance, and many mathematical patterns are not visual.

Tessellations are a powerful means of introducing (surreptitiously at first) every one of the themes listed above. They display pattern and order, a give a sense of harmony and balance, and a strong sense of symmetry: they show a wealth of relationships and tessellations are often analogous to one another and they always have an underlying structure.

The simplest tessellation is, plausibly, the square grid that can be used to cover the plane and which is the basis of co-ordinate geometry. Many more complicated tessellations have this grid as their 'skeleton' as in this example:

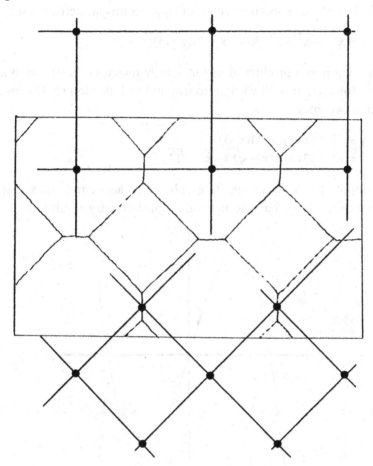

The skeleton of a tessellation

(1982a:19)

Patterns in algebra: the cubic

Patterns in algebra are less obvious because they are often hidden by the algebraic form rather than revealed. This is one reason why algebra is hard for most pupils: its patterns are highly abstract and difficult to appreciate. The quadratic equation,

$$x^2 - 6x - 7 = (x + 1)(x - 7)$$

does not immediately reveal to pupils its symmetry about x = 3 so they should be pleasantly surprised when they discover that the graph is a symmetrical parabola. This cubic equation appears unsymmetrical also:

$$y = x^3 - 6x^2 - x + 30 = (x + 2)(x - 3)(x - 5)$$

and writing it as a product of factors hardly increases the apparent symmetry. However, move the origin to its point of inflection (2, 12), and the equation becomes

$$y + 12 = (x+4)(x-1)(x-3)$$
or $$y = x^3 - 13x = x(x - \sqrt{13})(x - \sqrt{13})$$

This symmetry shows up on its graph which also emphasises that the original cubic always had geometrical point symmetry about (2,12):

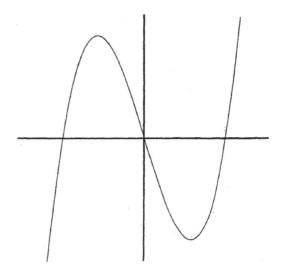

The symmetry of a cubic

This translation of the origin also illustrates invariance and unity-in-variety: the shape of the cubic has not changed and so in a sense all cubics of this shape, despite their different equations, are 'the same'.

The coefficient of x^2 is zero in this equation and so the sum of the roots is zero and will remain so if we subtract any linear function. Therefore if we draw a line intersecting the cubic in three points, the sums of the x co-ordinates will also be zero, an example of the interplay of algebraic and

Two parabola properties

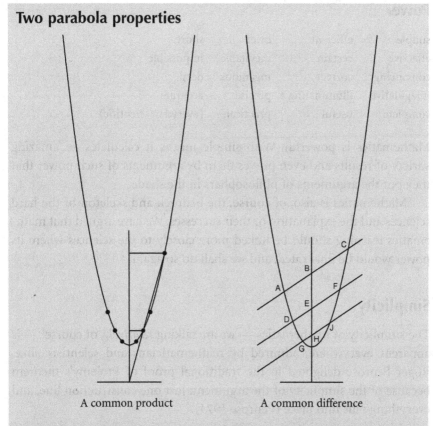

A common product   A common difference

We can deduce geometrical properties of the parabola from the sum and product of the two roots. The left-hand figure shows the parabola $y = x^2 + 10$ and the line $y = 7x$. Since the product of the roots is always 10, whatever the slope of the line, the products of the distances of the points of intersection from the y-axis are equal to 10, for any line $y = mx$. On the other hand, the sum of the solutions depends not on the function $x^2 + 10$ but only on the line. This translates into the property that if several parallel lines are drawn crossing the parabola, then $BC - AB = EF - DE = HJ - GH$ if all lengths are counted positive (1988c:117–118, 159–160).

geometrical features which can be used, for example, to prove the cubic version of Pascal's theorem.

This also illustrates the special role of the vertical line through the point of intersection which we need no longer think of as the y-axis of a graph but as a purely geometrical feature.

Power

simple	efficient	brief	short
effective	certain	inevitable	impossible
convincing	correct	ingenious	deep
imaginative	illuminating	precise	accurate
complete	useful	practical	(everyday/scientific)

Mathematics is powerful. With simple means it calculates an amazing variety of results and even proves them by arguments of such power that they put the arguments of philosophers in the shade.

Mathematics is also, of course, the bedrock and skeleton of the hard sciences and the explanation of their successes. We have argued that mathematics learning should be linked more closely to the sciences where its power would be illustrated, and we shall do so again.

Simplicity

The simplicity of mathematics — we are talking *relatively*, of course! — is apparent everywhere, admired by mathematicians and scientists alike. Roger Penrose delighted in the traditional proof of Ptolemy's theorem because of the simplicity of the argument: just one construction line, and everything falls into place (Penrose 1974).

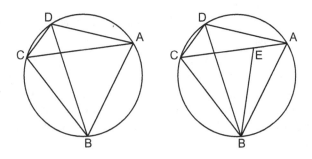

Ptolemy's theorem

The theorem states that AB.CD + BC.DA = AC.BD. The traditional construction is to draw the line CE so that the ∠DCE = ∠ACB. This seems at first sight both obscure and unmotivated: it is not on a closer examination because △AEB is similar to △DCB and △ABD to △EBC, and Ptolemy's conclusion follows by comparing ratios.

'See' the figure in the right way, a matter of perception not logic and what was obscure and difficult becomes plain and easy and pleasing.

Simplicity, unfortunately, is indeed relative and will not always be appreciated by pupils. Fortunately, there are other more obvious themes which will appeal to them.

Extremes as motivation

Extremes are extremely motivating, not least because they link themes of surprise and astonishment, and naked power. Very large numbers are extreme, and motivating. Primary pupils often get great pleasure by adding up extremely large numbers and the idea of very small numbers appeals too. You can talk about a millionth but can you find a millionth of anything? It is so extremely small!

In one particular classroom that I recall very well, the answer was, 'Yes!'. The classroom was on a second floor with enormous picture windows on two sides with wire-reinforced glass for safety's sake. So the question, 'Look around you! Can you see a million of anything?' was — eventually — answered when they realised that there were more than a million — countable — tiny squares in the wire-meshed glass panes.

There are other ways to illustrate a million. A large piece of graph paper (assembled by sellotaping together several smaller sheets) only 100 cm by 100 cm, divided into millimetre squares contains a million tiny squares. How long would it take to count them all at one second per square? How long would it really take to count them when you get to higher numbers and it takes several seconds to announce one number?

An historical story suggests another question for pupils: Archimedes wrote his *Sand Reckoner* to calculate how many grains of sand would fill the universe. Pupils might estimate the number of grains of sand to fill the classroom or how many grains of sand make a million.

Why is *The Guinness Book of Records* a best-seller? Because very large and very small numbers are fascinating and mysterious. How many people are there in the world? How do scientists measure the distance to the nearest stars? How do they know the size of an atom? How do they know what such figures mean?

Giant and tiny numbers

A2   How many seconds are there in a school day? How many seconds long is your life up to now?

A4   Fill a jar with sand. How many grains of sand in the jar? How big is a grain of sand? How much does a grain of sand weigh?

A5   What is the smallest thing you can weigh on kitchen scales? On scales you have designed yourself?

A6   How long would it take to count to a million?

B1   Johnny Jones has two parents, four grandparents, and so on. How many great-great-great … grandparents did he have 30 generations ago?

B2   The sun is about 1,392,000 km in diameter, and the earth is about 12,750 km diameter. If the sun is represented by an orange 10 centimetres across, how big would the earth be?

B3   How many mice make an elephant?

B4   You fold a piece of paper, then fold it again, and again … and again. How many times must it be folded before it is 1 metre thick?

(1987a:Unit 2:32–33)

Maxima and minima

> Problems concerned with … maxima and minima … are more attractive perhaps than other problems of comparable difficulty, and this may be due to a quite primitive reason.
>
> (Polya 1954:121)

Polya's 'primitive reason' was that we often have personal problems which involve greatest or least and so extremum problems *idealise* our everyday problems. Another even more primitive reason (which may involve a gender difference) is that small children are often fascinated by giant dinosaurs, very distant stars and galaxies and other extreme phenomena. Whatever the reason, maxima and minima problems are indeed especially intriguing.

Going to extremes

A1 A market gardener has to plant a large number of rose bushes. They must be as close together as possible but at least one metre from each other? How should they be arranged?

A3 What is the largest open box that can be cut out of a sheet 1 m square?

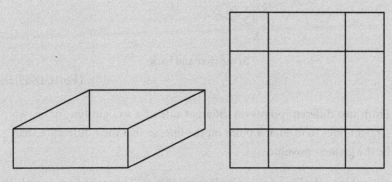

A maximum open box

A6 How large can the area of a triangle be if the perimeter is 30 cm?

B1 The sum of two numbers is 10. What is the largest their product can be?

B5 What is the smallest number with 18 different factors?

B6 Use the digits 1 to 9 to make two numbers, so that their product is as large as possible.

(1987a:Unit 42:304–305)

Maxima and minima are examples of extremes and many are very easy to state, *sometimes* easy to solve, and highly motivating as teachers have long realised — Robert Pott's *A Companion to Elementary Geometry* (1865), contained an entire chapter on extremal problems (Potts 1865:387–393). Here are some of his problems:

> From two given points on the same side of a straight line given in position, draw two straight lines which shall meet in that line, and make equal angles with it; also prove, that the sum of these two lines is less than the sum of any other two lines drawn to any other point on the line.

This is Pott's version of the 'shortest-route-to-the-seashore-and-back' puzzle!

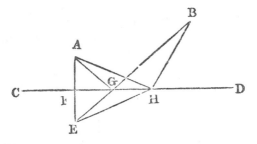

To the river and back

(Potts 1865:387)

From two different points on different side of a straight line, draw two straight lines to meet at a point on the line, so that their difference shall be the greatest possible.

This is a close relative of the previous problem. The next four puzzles are classics:

Of all triangles having the same base and equal vertical angles, the isosceles is the greatest.

Two pairs of equal straight lines being given, show how to construct with them the greatest parallelogram.

The perimeter of a square is less than that of any other parallelogram with equal area.

Inscribe the greatest parallelogram in the given semicircle.

Impossibility as a motivation

A surprising number of mathematical problems and theorems concern the impossible. It is an important and very motivating general concept that you can prove that *some things or states of affairs are impossible*.

'Right!', said Fred, 'I have thought of two numbers, their sum is 10 and their product is 23. What numbers did I think of?' 'You didn't,' replied Mary, 'two numbers like that don't exist!' 'Yes, they do,' insisted Fred. Which of them is correct?

Questions of existence

A   Is there a number which you can multiply by 5 to get 7?

B   What is the largest number less than 10?

D   Is there a polyhedron with seven triangular faces?

F   Does every differential equation have a solution?

G   Is it possible to find a point which is equally distant from all four vertices of this quadrilateral?

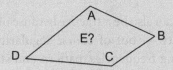

Is there an equidistant point?

H   For the same quadrilateral, is it possible to find four points, one on each side, which are the vertices of a parallelogram?

L   Does a set of points in the plane have an average?

N   Is there a largest number which is *not* the sum of three squares?

O   Do oval shapes have diameters?

Q   Is it possible to fit the 12 pentominoes into a rectangle 3 by 20?

Many puzzles can be out in the form of questions about what is or what might possible be.

Expectations met and broken

Order and pattern meet our expectations and we are both surprised and curious when our expectations are dashed. Strangely, expectations met AND expectations dashed can be attractive. We are satisfied when our expectations are met but may be more excited when our predictions are confounded.

Counterexamples create surprise by contradicting our expectations. Pattern is attractive but when a pattern is broken we naturally start to *wonder* — the source of all art and science according to Einstein. The simple is attractive but so is the complicated provided we can find some sort of order in it. Mystery is attractive but so is revelation!

Naturally, expectations are often broken in science:

> Until recently, traditional scientific thinking held that ultimately every-
> thing in the universe has a symmetrical equal and opposite counterpart ...
> but a beautiful theory has led to the finding that universal symmetry ...
> is sometimes slightly violated, and scientists believe they have come upon
> a truth ever more beautiful than the simple perfection of symmetry.
>
> (1980–1983, Newsletter #7, 'Beauty')

Strangely, randomness can also be attractive because of the apparent absence of any of any regularity — but of course regularity appears *on average* when you start counting frequencies, pattern and structure appear when you calculate the probabilities of, say, throwing a ten with three dice by enumerating all the possible combinations.

(The motivation of giant numbers appear here also: how many ways are there to arrange the 30 pupils in class in a straight line? Most pupils initially find the answer incredible.)

Expectation and intuition

However, *before* pupils can be surprised or their expectations broken, they must develop some feeling, some familiarity, with the situation in question. Young pupils who achieve the goal of 'being friends with the numbers 1 to 100' will be readier to think about them naturally, to ask questions and to wonder, than pupils who have no such familiarity, to whom numbers remain strange and confusing.

Idiosyncratic differences in aesthetic appreciation

Krutetskii (1976) noted that appreciating the elegance of a solution was typical of the more capable students. It does not follow that all pupils cannot appreciate aspects of mathematics aesthetically though it may be that certain components of mathematical beauty will be appreciated more readily by most pupils, while others will mean most to the stronger. According to

Freeman Dyson, a brilliant young mathematician who long ago became a world-class physicist:

> Unfashionable mathematics is mainly concerned with things of accidental beauty, special functions, particular number fields, exceptional algebras, sporadic finite groups. It is among these unorganised and undisciplined parts of mathematics that I would advise you to look for the next revolution in physics. They have a quality of strangeness, of unexpectedness. They do not fit easily into the smooth logical structures of Bourbaki. Just for that reason we should cherish and cultivate them, remembering the words of Francis Bacon, 'There is no beauty that hath not some strangeness in the proportion'.
>
> (Freeman Dyson 1983:47)

The general and universal is attractive but so also is the idiosyncratic and the contingent, though possibly to different mathematicians and to different pupils. The stronger pupils will find it easier to appreciate the general, especially the more abstract, features which cause the most difficulty to the majority who need the motivation of the contingent and the concrete, the unusual, the striking and the strange, even as they work towards a better understanding of more difficult and advanced ideas.

6

Proof in School Classrooms

Is there a place for proof in school mathematics today? (...) Whatever we conclude about the introduction of proof structure and reasoning into the upper part of the secondary school it must be clear from the outset that there can be no question of introducing formal or informal (sic) proofs into primary mathematics.

(Baron 1973:200–207)

Proof — that is arguments which make you feel sure, which produce insight, which make you jump from your seat, or sit back with the feeling that you've cracked it for sure — are intimately linked to motivation and aesthetic judgement. If proof is missed out, pupils miss out on one of the chief rewards of doing mathematics.

(1995c:24–25)

A 'theorem' is simply a mathematical statement *of some interest* which is known to be true because it has a proof, i.e. a logical demonstration of the truth of that statement.

(Royal Society/JMC 2001:57)

Which of these claims is correct?

Proof is a central concept in mathematics: indeed, it is the feature that distinguishes mathematics (and abstract games) from science, which proves nothing, though scientists often claim, for example, that they have

115

proved that the background radiation they have detected *really is* from the Big Bang. No, scientists can never escape from the possibility that tomorrow evidence will be discovered to undermine their pet theory, even if it does not completely demolish it.

Unfortunately, because of historical links with Euclidean geometry, proof has been regarded over the years as frankly rather difficult and therefore beyond most pupils and as a 'hard' aspect of mathematics to be approached with care and even trepidation by both pupils and teachers. This is nonsense.

Instead of seeing proof in the context of professional mathematics and in the history of mathematics — where Euclid is a shining beacon — which means starting with preconceived ideas of what a 'proof' should look like and the form it should take, teachers should look at the way pupils behave, at their thought processes, at their arguments and then pick out and highlight those arguments that are sound *because a proof is only a sound argument* — usually of something that is judged to be significant so that a sound argument that $12 \times 15 = 180$ is not dignified with the label 'proof' though logically speaking it is just that.

The italics in the third epigraph are ours: 'of some interest' is an important qualification but also a vague one. What interests a primary pupil may be taken for granted by a secondary pupil and be positively boring for a professional mathematician: but we are not teaching the latter. And is it the statement itself which is interesting — to whom? — or the pupil who is interested?

Proof exists at every level of difficulty in mathematics so it is not limited to stronger or older pupils but should be a familiar idea to all pupils beyond the very youngest. Of course, not all *claimed* proofs are sound and pupils will often produce unconvincing 'proofs' or proofs that contain obvious errors but they will also produce argument-proofs that are completely sound, both to them and to the teacher.

The omission of proof is a perfect example of *false simplicity*. By playing down what is mistakenly conceived to be a difficult aspect of mathematics, pupils are deprived of all the satisfactions of searching for good arguments, and then getting a 'kick' (Hardy's term) or 'buzz' from finding them, plus the aesthetic satisfaction that comes from being convinced that something is indeed the case.

The result of this mistaken strategy is to create *false simplicity* while depriving pupils of *rich complexity*. About the time of the Cockcroft Report, a report from the Stoke Rochford Conference, noted that it was possible to go through O-level and A-level 'without meeting the idea of proof or feeling the need for proof' and also noted that 'concern was expressed over the decline in the ability to appreciate the nature of proof which seemed to be observable among A-level students' (SMP 1980).

A little later, *School Mathematics in the 1990s* (Howson & Wilson 1987:61) considered the possibility that 'one abandons the idea that geometry should/can be treated in the school as a system of knowledge (organised deductively or not) … ' and concluded that if that step were taken then, 'Consequences: 3: The teaching of 'proof' at school level will be severely restricted.'

Really? It will be somewhat restricted, because traditional Euclid is a fine source of provable propositions (and fascinating experiments) but proof is available to pupils in any area of mathematics whatsoever — provided teachers 'see' that it is present. It is not absence of opportunity that is the problem, but the inability to see the many opportunities there are — often, as the Howson and Wilson conclusion suggests, because of a fixation on traditional ideas of proof in Euclid (1995c:24–25).

Textbooks today demotivate because they omit proof as it is 'too difficult' but then find that they have little that is challenging or attractive to present to pupils except some very basic ideas and facts with which the pupil can 'do' nothing because he or she lacks intuition, lacks familiarity, lacks background and lacks appreciation.

The opportunities are there in the National Curriculum. As Melissa Rodd and John Monaghan point out, discussing 'school mathematics and mathematical proof':

Infants, in Key Stage 1, are encouraged to explain their methods in their number work.

The Key Stage 2 Programme of Study links this requirement for pupils to explain their reasoning with the admonishment to 'develop logical thinking'.

In Key Stage 3 these significant terms are used: generalisation, conjecture, counterexample, step-by-step deduction, practical demonstration, assumptions, proof.

In the Key Stage 4 higher programme, Key Stage 3 proof concepts are increased with pupils being expected to do 'short chains of deductive reasoning'.

(Rodd & Monaghan in Haggarty 2002:Ch. 5)

On the other hand, according to the current National Curriculum (2007), at Level 5 pupils will 'draw simple conclusions of their own and give an explanation of their reasoning' (Attainment target 1: Using and applying mathematics) but most 11 year olds are only expected to reach Level 4.

Most 14 year olds are expected to reach Levels 5 or 6 but at Level 6 'pupils are beginning (sic) to give mathematical justifications' and so many 14 year olds are expected to miss out. Level 7 at which pupils 'justify their generalisations, arguments or solutions' and 'appreciate the difference between mathematical explanation and experimental evidence' is apparently beyond most 14 year olds.

This is absurd and the consequences for proof in school disastrous.

The Justifying and Proving in School Mathematics project started in 1995 and was designed study the achievement in the field of proof of 'high-attaining year 10 students'. Its first finding was: 'High-attaining year 10 students show a consistent pattern of poor performance in constructing proofs' (Healy & Hoyles 1998:2). This is hardly surprising if they have not been creating proofs since they were much younger so their past experience is minimal:

Empirical verification was the most popular form of argument (sic) used by students in their attempts to construct proofs, and in problems where empirical examples were not easily generated, the majority of students were unable to engage in the process of proving. (*Ibid.*:2)

Their 'major finding' was that,

Most high-attaining Year 10 students after following the National Curriculum for 6 years are unable to distinguish and describe mathematical properties relevant to a proof and use deductive reasoning in their arguments. (*Ibid.*:6)

Since proof is treated by the National Curriculum as a difficult topic to be introduced very late in the day, their conclusion is expected. If the NC emphasised argument more, from the very start, then the result, we might confidently suppose, would be very different.

The many functions of proof

Some people believe that a theorem is proved when a logically correct proof is given: but some people believe that a theorem is proved only when the student sees why it is inevitably true. The author tends to belong to this second school of thought.

(Hamming 1980:155)

Pupils at first tend to find the second type of proof more convincing. They will be most impressed by their very own 'Aha!' penny-dropping subjective experiences. It takes more sophistication to appreciate that a logical argument where no penny drops, can still be totally convincing *and even worthwhile.*

Superficially, proofs might seem to have a single essential function, that of providing certainty, of absolutely confirming what we suspect, but that is false. That is only one function among several. Yes, we often suspect that a result is true and would like to confirm it and to persuade others — the teacher and the other pupils. However, we are often completely convinced of a result — by geometrical drawing, by calculation, or by spotting a convincing pattern — in which case what is the function of proof? Why did Gauss produce six proofs of the law of quadratic reciprocity with a seventh found among his papers? Was the first proof suspect? The second, the third? Of course not: Gauss was not searching for certainty but for *insight* and *understanding.*

Proofs do provide certainty — if you are confident that your assumptions and your arguments are sound — and inferior proofs do no more than that: superior proofs rely on imagination and insight to *illuminate* the problem so that you end up not only convinced that what you want to prove is indeed true — illuminating proofs of difficult propositions are incomparably the *most* convincing — but you understand it much better!

Gila Hanna and Ed Barbeau on the functions of proof

1   Verification. Validating correctness.
2   Explanation. Answering the question 'Why?'
3   Conviction. Removing doubt.
4   Systematization. Fitting mathematical results into a wider context.
5   Discovery. Inventing new results.
6   Communication. Transmitting (sic) mathematical knowledge and under-
    standing.
7   Enjoyment. Meeting an intellectual challenge elegantly.

(Hanna & Barbeau 2008)

We might say that proofs need ideas, ideas depend on imagination and imagination needs intuition, so proofs beyond the trivial and routine force you to explore the mathematical world more deeply and it is what you discover on your exploration that gives the proof a far greater value than merely confirming a 'fact'.

The best proofs are always illuminating: the worst and the least satisfying confirm the claimed result, but otherwise leave us in total darkness. Such proofs are, as it were, dead ends. They lead nowhere, unlike the illuminating proof which immediately suggests new ideas, new directions and new possibilities — so the most famous proofs are all brilliantly illuminating. Morris Kline commented (Kline 1973) that,

> Much research on new proofs of theorems already correctly established is undertaken simply because the existing proofs have no aesthetic appeal,

which is to say, they are unilluminating because illumination is *always* appealing. With new ideas come new methods, new techniques, and *new approaches*, so that the effects of a novel proof spread outwards far beyond the original problem:

> Shishijura's method of attack has enabled mathematicians to get deeper into the detail of the complex dynamics associated with the Mandelbrot set. 'The importance of this is not the result itself, but the new techniques introduced.'

(Brown 1991:22)

The problem for teachers should be, not whether the argument-proof would impress a professional mathematician but whether the pupil is justified *at their level* in appreciating the argument as a proof.

Teachers should not impose their ideas of proofs on pupils, but look instead for examples of thinking in children's problem solving work that has the characteristics of proof, and then interpret and explicitly label such thinking as proof wherever it occurs.

In particular, teachers must stop thinking of proofs as necessarily long with many individual steps, and accept that even the shortest chain of reasoning is a proof provided it displays the usual features such as clarity, conviction and communication.

It is also necessary to stop associating proof with particular parts of mathematics, for example with geometry.

Calculation, technique and proof

Is a calculation a theorem or a construction? It is both — it 'proves' that the final answer is correct, even as it constructs it. Any arithmetical calculation — to start with the simplest possibilities — is a proof if we ignore the vague requirement that it be 'of some interest'. This sum,

$$
\begin{array}{r}
173 \\
\underline{24} \\
692 \\
\underline{3460} \\
4152
\end{array}
$$

proves that 173×24 equals 4152. If we break down the formal and routine sum and write it out at length, the fact that it is indeed a proof is even more obvious:

$$
\begin{aligned}
173 \times 24 &= 173 \times (20 + 4) = 173 \times 20 + 173 \times 4 \\
&= (2000 + 1400 + 60) + (400 + 280 + 12) \\
&= 3460 + 692 \\
&= 4152
\end{aligned}
$$

This is a chain of logically sound inferences, leading up to a logically sound conclusion, that is, a proof, albeit of little interest. However, even in this 'trivial' example we can add interest by rephrasing the instruction from 'Calculate 173×24' to 'Prove that 173×24 is 4152' or 'Prove in as many ways as you can that 173×24 is 4152'.

The fact that the same answer can be reached in more than one way emphasises the idea that any one conclusion typically has many proofs.

Calculators of course make arithmetic with integers and decimals absurdly 'easy' and when used too often — as they are by many children — they handicap pupils by undermining arithmetic fluency. However, the tables can be turned if pupils are allowed to find the answer by the calculator and then asked to prove that the answer is correct.

Very simple or routine arguments are usually not labelled proofs because they are not considered sufficiently 'serious' and lack 'interest'. The simplest calculations and algebraic manipulations are indeed easy and nothing to write home about — for the experienced student — but that is a social not a logical point and if we are to induct pupils into the idea of proof and its distinctive role in mathematics, then we should focus on both aspects. By all means — it's a general concept! — highlight cunning proofs which are surprising — powerful — difficult — whatever — but also give credit to pupils for simple arguments at their level of achievement that are logically sound and warrant the label of 'proof'.

Many manipulations, some of them elementary, deserve the title of proof, including algebraic identities such as these:

$$x^3 + y^3 = (x + y)(x^2 - xy + y^2)$$
$$x^4 + 4y^4 = (x^2 - 2xy + 2y^2)(x^2 + 2xy + 2y^2)$$

A proof, by direct calculation, of the second is especially surprising and therefore especially deserving of the label because $x^4 + 4y^4$ looks so much like $x^2 + 4y^2$ which has no algebraic factors.

Quadratic equations offer many opportunities for arguments/proofs but mostly at a level which only the stronger pupils can be expected to appreciate because they depend on a fluency in algebraic manipulation plus *interpretation of the results*. In the box below is a simple argument about quadratics which is easier for many pupils. A much more sophisticated argument that requires greater ability in manipulation and interpretation goes like this:

Suppose that p and q are two roots of the equation $x^2 + A = Bx$

Then, by definition, $\qquad p^2 + A = Bp$

and $\qquad\qquad\qquad\qquad q^2 + A = Bq$

and so $\qquad\qquad\qquad p^2 - q^2 = Bp - Bq$

and $\qquad\qquad\qquad (p - q)(p + q) = B(p - q)$

Therefore, if the roots are distinct, their sum is B. We can also argue that if $p^2 + A = Bp$ then $Bp - p^2 = A$ and $p(B-p) = A$, so recovering the idea that there are two numbers, p and B-p, whose sum is B and whose product is A. By substitution, if p is a root then B-p is a root also.

Alternatively, if $\qquad\qquad p^2 + A = Bp$

then $\qquad\qquad\qquad\qquad p + A/p = B$

which also on interpretation says that there are two numbers, p and A/p whose product is obviously A and whose sum is B. Once again, by substitution, if p is a root then A/p is a root.

Quadratic equations and proof

Find both solutions of the equation: $T^2 - 6T + 4 = 0$ by completing this argument: 'The sum of the roots is 6. Therefore the average of the two roots is 3, and the roots can be written in the form 3+p and 3−p. But the product of the roots is 4. Therefore …'

(1980–1983, #2:11)

Pupils can very often discover by experiment that if an equation such as $p^2 + A = Bp$ does have two solutions, then their product is A and their sum B (raising the puzzle: what if the equation seems not to have two solutions?). This leads to a logical argument or proof which is much easier for most pupils that the usual 'completing the square': if the two solutions sum to B then they can be thought of as ½B − e and ½B + e in which case their product,

$$(\tfrac{1}{2}B - e)(\tfrac{1}{2}B - e) = A \text{ or } \tfrac{1}{4}B^2 - e^2 = A$$

And e can easily be calculated. In the example, $p^2 + 21 = 10p$, and the solutions must be 5+e and 5−e, so $25 - e^2 = 21$ and e is 2. (Notice that this argument works for pupils who have not yet met negative numbers.) This argument can also be used to motivate the idea that if e^2 is not a perfect square, there may still be solutions.

(1980–1983, #8:28–29)(1988c:114)

Proof, intuition and mathematical recreations

The arguments that make up a proof are seldom totally obvious, and the harder they are to find the more it is likely that we shall not be able to say exactly 'where they came from'. In other words, proofs often depend in the final resort on the imagination and *intuition* of the student. As we have seen, however, intuition is difficult to develop and takes much time and experience — but the results are overwhelmingly superior to the fast and shallow learning which characterises most mathematics classrooms.

Therefore, we can maximise the chances of pupils being able to prove speculative results (and think of such potential theorems in the first place) if we put them in situations where their intuition is relatively strong. Mathematical recreations are a good place to start. The next two examples are well-known and provide many opportunities for simple (and not-so-simple) arguments. They also give an advantage, it must be admitted, to pupils who are visually adept. The other side of that coin is that they provide practice for those who are less visually minded and who will benefit from developing their visual sense.

Polyominoes

Polyominoes are ingenious, simple, and enjoyable to play with — and highly tactile if the pupils have actual tiles — and potentially challenging because their shapes are quite weird and do not easily fit together. They offer opportunities for proof by many methods, notably:

- Simple trial and error.
- Enumeration of all the possibilities.
- Exploiting the properties of corners and straight edges, which limit the possible arrangements of the polyominoes.
- Colouring arguments, usually based on colouring each polyominoes in a chessboard pattern of two-colour squares.
- Constructing basic assemblies which can then be repeated.
- Developing specific tactics and more general strategies.

All the puzzles in the box are open to simple and convincing proofs in everyday English.

Polyomino puzzles

Prove that there are only five tetrominoes.

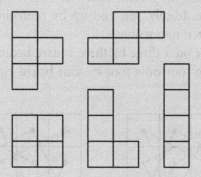

The five tetrominoes

Prove that copies of each individual tetromino can be used to tile the infinite plane.

Prove that copies of each individual tetromino can be used to tile the infinite plane in many different ways. (In how many different ways?)

Prove that although the total area of the five tetrominoes is 20 units, they cannot be fitted together inside a four by five rectangle.

Prove that tetrominoes can be used to tile this endless strip:

Can tetrominoes tile this strip?

Can copies of the T-tetromino be used to tile the plane?

Which tetrominoes have the smallest circumerence? The largest?

How many pentominoes are there?

What is the smallest polyomino that contains a hole?

Knight tours

The goal of a knight tour is to start with a chess knight on one square of a board, and by moving the knight visit every other square on the board once and only once. Ideally you end up by returning to your starting square but this is often impossible.

There is no tour on a three by three square because no square leads to the centre and no tour on a four by four board either. (How can this be proved?)

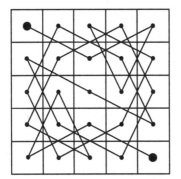 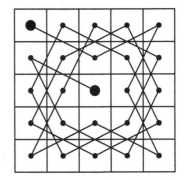

Knight tours in a five by five board

There are many solutions on a five by five board, including the symmetrical solution on the left which is hard to find and the easier to find solution on the right which starts in a corner and keeps as close as possible to the edges until it finally leaps into the central cell. In none of these five by five solutions, however, is the final cell a knight move from the starting cell. This is inevitable, and can also be proved (2007:12–14).

Proof by looking

There are many opportunities for proofs-by-looking or 'proofs without words' in elementary mathematics. Pupils can then be challenged to explain in words what they are doing, especially when generalising from a particular figure to a general case.

The next arrangement shows that $5 \times 7 = 7 \times 5$. Yes, it is simple, but it is also totally convincing and it is a proof of a fact that, until pupils are accustomed to it, is by no means obvious.

It also illustrates a deep principle of combinatorics: you can obtain many striking results by simply counting one arrangement of objects in several different ways, as the next figure illustrates.

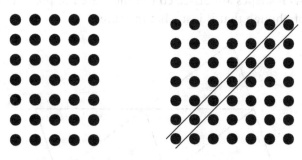

A proof that $5 \times 7 = 7 \times 5$ A formula for triangular numbers

By dividing the right-hand square arrangement of dots in different ways we prove that $T_n + (n+1) + T_n = (n+1)^2$, where T_n is the nth triangular number.

The next figure, which can be constructed with Cuisennaire rods, shows that $1 + 2 + \ldots + 7$ repeated makes a rectangle 7×8. You can 'see' that the pattern will work for any choice of numbers 1 to N just as confidently as you can *see* that any algebraic proof works.

Similar diagrams are often used to *explain*, for example, why $(x + a)(x + b) = x^2 + (a+b)x + ab$ but the figure is also a *proof-by-looking* that this is indeed so.

Finally, the right-hand figure demonstrates that the sum of the first n cubes is the square of the nth triangular number:

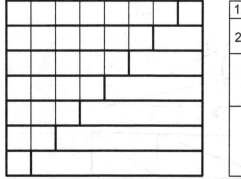 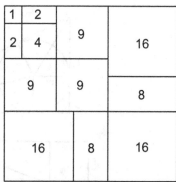

A proof with Cuisennaire rods   Summing the cubes

Proof by dissection and assembly

The formulae for the areas of basic shapes are usually explained by diagrams in which shapes are dissected but these are not presented as examples of proof, though that is just what they are:

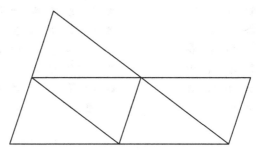

Dissecting a triangle into a parallelogram

Pupils who know that you can dissect a triangle into four identical small triangles *and* who know the properties of angles and parallel lines can *infer* logically that if you take the top triangle and move it as in the figure, the result is a parallelogram — of equal area.

A Chinese proof by dissection

What is the radius of the incircle of this right-angled triangle? Liu Hui in his third century book, *Commentary on The Nine Chapters of the Mathematical Art*, drew this figure (we have simplified it slightly):

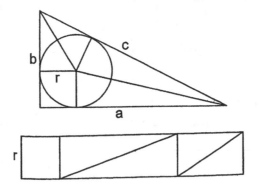

Proving the radius of the incircle by dissection

The height of the rectangle is r, and its length is $1/2(a + b + c)$. Comparing areas, $r = ab/(a + b + c)$ (Yan & Shiran 1987:70–71).

From 2D to 3D and back

Calissons are French sweets in the shape of two adjoining equilateral triangles. They are sold in a hexagonal box and it seems by experience that one third of calissons in any arrangement face in each of the three possible directions. Why should this be?

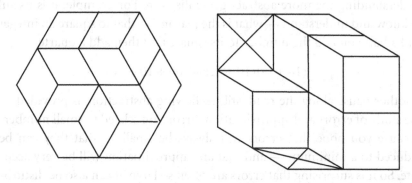

The puzzle of the calissons

This is the spectacularly visual solution: the box has been rotated and distorted to highlight the image as a picture of four cubes (and possible a fifth hidden cube) inside a $2 \times 2 \times 2$ cubical box. The 'faces' marked with crosses are two faces of the $2 \times 2 \times 2$ cube. Looking at the large cube from each different direction we naturally see $2 \times 2 = 4$ faces, corresponding to the four calissons facing in each of the three directions (David & Tomei 1989).

Proto-algebra: proof by looking-at-calculations

The premature use of algebra, either in algorithms or pseudo-algorithms, has the same effect as the premature use of other methods that pre-empt understanding, by allowing pupils to obtain solutions with little or no understanding of the logic of the problem.

Proto-algebra — generalising from arithmetical calculations — is one way of *slowing down* their understanding and preventing them from leaping to simple-minded naive 'understanding' too quickly and easily. Pupils who can calculate mentally that,

$$14 \times 17 = 14 \times 10 + 14 \times 7 = 100 + 40 + 70 + 28 = 238$$

are also proving-by-calculation — and they are more than half way to 'proving meaning seeing' that once again:

$$(x + a)(x + b) = x^2 + (a+b)x + ab$$

Pupils can take other steps that will slow down their progress to really understanding the more abstract generalisation. For example, it is useful to know and understand — what is the reason? — that to square an integer and a half you add the number to its square and then add a quarter:

$$(10 + \tfrac{1}{2})(10 + \tfrac{1}{2}) = 100 + 10 + \tfrac{1}{4}$$

Another pause along the road, still eschewing abstraction, is provided by the study of errors and approximation. Errors are related to small numbers because you hope that errors will always be small — that they can be reduced to a minimum — and that any approximation will be very accurate. So it is surprising that errors are often so large. It can also be disturbing and provoking because *error* has very negative emotional connotations whereas *approximation* and *estimation* are emotionally neutral.

Questions such as these should be extremely thought provoking and can lead in many directions:

B5   Is 34 cubed a good approximation to 34.7 cubed?

C2   Two numbers, call them P and Q, are rounded down to the nearest whole number before being multiplied together. What is the greatest possible error due to the rounding down?

(1987a:Unit 15)

Series also provide great scope for game-like arithmetic and arguments, as well as being attractive in themselves because they produce so many patterns and so many *surprises*.

There are many ways to argue that it is reasonable to say that,

$$1/2 + 1/4 + 1/8 + 1/16 + 1/32 + 1/64 + \dots \text{ 'equals' } 1$$

One solution, at the level of proto-algebra, is to calculate the partial sums, which turn out to be 1/2, 3/4, 7/8, 15/16, 31/32 … and so on. This arithmetical argument is also convincing but more sophisticated: we replace 1/2 by 1 − 1/2 and 1/4 by 1/2 − 1/4 and so on, and the series becomes, by a plausible argument:

$$(1 - 1/2) + (1/2 - 1/4) + (1/4 - 1/8) + (1/8 - 1/16) + \ldots = 1$$

The properties of the Fibonacci sequence provide opportunities to argue correctly, to draw cunning conclusions and to *prove*, long before any matching results can be expressed in algebra. Thus, the product of two consecutive terms,

$$
\begin{aligned}
55 &\times 34 \\
&= (34 + 21) \times 34 \\
&= 34^2 + 34 \times 21 \\
&= 34^2 + 21^2 + 21 \times 13 \\
&= \ldots \\
&= 34^2 + 21^2 + 13^2 + 8^2 + 5^2 + 3^2 + 2^2 + 1^2 + 1^2
\end{aligned}
$$

Moreover, it is 'easy to infer' that this pattern works for any pair of starting numbers, because it is so strong. So, the sum of the squares of successive Fibonacci terms from 1, is the product of two consecutive terms.

In all these examples of proto-algebra, algebraic literals are absent, replaced by actual numbers as *representative* of the sought pattern — as was done by Diophantus and other early algebraists.

The aesthetics of proofs

Proofs have another invaluable feature: they involve and are judged by much the same aesthetic criteria as mathematical objects, ideas and methods.

Proofs can be valued because they are simple, short or surprising; because they are powerful and can be used with modifications to prove many other propositions; or illuminating and deep depending on novel insights; or because they cunningly exploit patterns, symmetries, hidden connections, subtle analogies or underlying structures … and so on.

Challenged to prove that $111111111 = 3^2 \times 37 \times 333667$ some pupils will be interested because they are surprised, because the pattern is unexpected, others simply because they like a challenge and many pupils will learn something by looking for a brief and simple proof. As it so happens,

$$333667 \times 3 = 1001001 \text{ and } 3 \times 37 = 111$$

These observations not only make the proof as simple as possible, but they invite questions about the strange numbers in the puzzle: 333667, 37 and 111111111. Here are three pertinent observations:

$$11\,111\,111\,111 = 21649 \times 513239$$

and is therefore extraordinarily difficult to factorise[‡] and

$$33667 \times 3 = 101001 \quad \text{and} \quad 36667 \times 3 = 110001$$

Naturally proofs give pupils a 'kick', to use Hardy's term. You get a 'buzz' or 'kick' from knowing that you have cracked the problem and understood it. You can also get a 'kick' out of knowing that you proved what you previously only suspected — even if your suspicions were very strong and your confidence cast-iron.

Omit proof, and you leave out one of the greatest motivating factors that the teacher possesses! Provided, that is, that you understand 'proof' as the child producing sound arguments *rather than* aping the textbook or traditional Euclidean theorems.

Proof of impossibility

A powerful source of aesthetic reward is the idea that something is impossible and that you can prove this. Professional mathematicians themselves are intrigued by the very idea that something *cannot exist* — whether it is an odd perfect number, a closed algebraic formula for solving quintics, or a solution to Fermat's Last Theorem.

Pupils can also get a kick out of the idea that some mathematical things which you can talk about all right — that's the curious feature! — like mermaids and griffins, don't exist. We can talk about them, we can argue about them, we can draw conclusions about them — but when you get to the crunch they aren't 'there', anywhere.

It has been claimed that most children leant to say 'no' before they say 'yes' — simply because negation is a stronger expression of feeling than

---

[‡]You could use (by rote!) a site such as Dario Alpern's Factorization using the Elliptic Curve Method at http://www.alpetron.com.ar/ECM/HTM.

affirmation. Perhaps the same idea applies to ideas of proof. It is not obvious why it should be necessary to prove facts that experience plainly teaches so the Greeks are always given credit for realising that even very 'simple' facts can be proved from even 'simpler' axioms. In contrast, when something goes wrong it is quite natural to try to wonder why.

Anyway, here is a typical 13-year-old pupil, Mary Watson, explaining why an apparently simple puzzle — which appears to have no solution after repeated trial and error — really *cannot possibly* have a solution. The puzzle is to fill in the four cells to give the sums marked on each side:

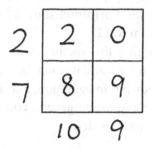

You cannot fill the cells to match the sums

> This problem cannot be solved because if you take the line which has the number 2 on the highest number you could have would be 2 so if you put 2 under the ten and 0 under the 9 that would be 2 but if you put the 9 beside the 7 and the 8 which would be left over from the 10 under the 10 it would not work out. 7 would add up to 17.

(1980–1983, #6 Newsletter:13)

Mary's friend Jennifer Dunham also proved that the puzzle is impossible, and then explained her answer to a slightly different puzzle:

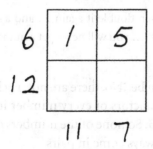

Different sums, different solutions

Sir gave me this problem where I have to find numbers that add up to the numbers outside the square. I found out for example that every number that is used to make six, ex. $1 + 5 = 6$, $1/2 + 5\ 1/2 = 6$. I used the numbers in the column and all that I had to do was to find other numbers that fit to make up that number, this can go in sequence, I'd know what numbers other squares have by this idea, it is complicating, but easy when you get to do it. (1986a:16)

Jennifer is describing, as a 13 year old never before asked to write any proof, how you can pick any pair of numbers summing to 6, fill the remaining cells to make the horizontal sums 11 and 7, and the second vertical row will always sum to 12.

She does not realise that this method will work for any choice of $a + b = 6$ *because* $11 + 7 = 6 + 12$, and it would be astonishing if she did. But she did 'have an idea', showing imagination, and it worked: a great start. Her solution involves two important general concepts, in this case about equations, but applicable more widely:

- Some problems have any number of solutions.
- There can be a rule or method for finding all the solutions.

To an adult teacher this is quite a 'primitive' attempt, but it was also the first time that these pupils had ever been asked to *prove* anything, and as first attempts go they are excellent and very promising. (Pupils who are unable to explain clearly in writing may be able to produce much clearer explanations verbally: if the teacher has a tape recorder these can then be transcribed, within reason.)

Jennifer Dunham also thought of this problem (published in *The Problem Solver*, 2:13) while thinking about something quite different:

If you double 6 and then double it again … and again … and again … like this: 6, 12, 24, 48, 96 … you will never get a number which is a perfect square.

Why not? Let the solution be, 'No, there are not. Each number is double the previous number, so the factors of every number in the sequence are several 2s and one factor of 3. So none of the numbers can be a square because the factors of a square always come in pairs.'

I will confidently assert that that solution is a proof at the pupil's level and is a proof moreover which may well be produced in this or some equivalent form by a pupil (1987b).

Proof and generality

It is an important general concept that some proofs can be generalised so that they work for any example of a *type* of problem. They work for a specific problem to start with, but it turns out that they work for a lot of other problems, maybe with a bit of adjustment. Such proofs turn out to be *methods*: every problem of the same *type*, can be solved using the same procedure.

Hande and Selina's proofs invite a question: can they be generalised? Can identical or very similar arguments be used about other problems? Proofs can be very general or very specific.

Pupils will naturally start with more specific propositions proved with relatively specific arguments as Hande and Selina did in the proofs just quoted (not least because many simple conclusions can be proved without using algebra and later proved more generally with algebra).

Pupils can then be expected to move on to harder proofs — harder *because* they are more general — bearing in mind that a slight increase in generality may take the pupil much further up the steep curve of difficulty which we have already discussed.

A small handful of potential proofs for pupils

- Every prime except 2 and 3 is $6n \pm 1$.
- There is no tessellation of regular pentagons that does not contain gaps or overlaps.
- How many repeating tessellations are there of equilateral triangles and squares?
- How do you prove that $\sqrt{5}$ as a decimal starts with the digits 2.236...
- Prove that 1009 is a prime number.
- Prove that 8357 and 4096 have no common factor.
- Prove that 1 less than a square is never a prime number, except for 3.

More examples of proofs by pupils

The following proofs, though simple, show several important features:

- They were not done quickly. In particular, the proof of the billiard ball problem took hours of investigation before the pupil had his particular insight. Pupils, like professionals, often need time to succeed at mathematics, time which they are very seldom allowed.
- They create personal conviction; this is crucial and it is often enough. If I prove something by elementary algebra then the chances are that, after checking carefully, I shall be so strongly convinced that I feel no need to ask someone else to check my proof as a precaution. (I may want to show off my conclusion, but that is another matter.)
- They require imagination. It is a common misconception that 'proof' means 'only' being terribly logical. It does. The mistake is to assume that you can be logical *without using imagination*. This is seldom possible, because to be logical you have to make connections — often distant and obscure connections — and so the most powerful logical arguments — including the most famous proofs by the greatest mathematicians — are without exception wonderfully imaginative. Hence Voltaire's saying, maybe unfair to the poet, that,

> There was far more imagination in the head of Archimedes than in that of Homer.
>
> (Voltaire 1764:#3)

All the proofs require very little calculation, with the exception of Daniel Schulz's proof (pages 148–9) which has lots of calculation which is, however, quite routine once you understand how to use weighted means. In the other proofs, minimal calculation is needed to support the feature already mentioned — imagination.

The proofs themselves, and the ideas and methods used, all suggest further questions and problems.

A very simple game

Like the three by three magic square this game can be described as a recreation for small children but that does not mean that it is not mathematical and extremely instructive, for the pupils and for us. The following account is reproduced from *Investigator 12* (March 1988), the magazine of the SMILE Project.

The game of Corner

S(hirley) C(larke): Jermaine and Nicholas, fourth year juniors unused to problem solving and investigational work, tried out Corner.

This game is played with just one castle on a chess board. The first player puts the castle on an edge square of his or her choice. The two players then take turns to move the castle as many squares to the left as they like, or as many squares down. The winner is the player who moves the castle onto the winning lower-left corner square.

> I asked them to see if they could find strategies for winning, and to record all their thoughts, discussions and findings. After two or three games, Jermaine noticed that the player who placed the castle first seemed to win. After discussion with me about how to be sure of this, they decided 12 games would be enough of a test. They started their test, playing each game with no apparent strategy other than alternating who went first. They kept a record of their results. We had marked the chessboard A to H along the top and 1 to 8 along the side. After eight games had been played, Nicholas announced that if you got the castle on C6 you always won. I asked him to explain why. He showed me, moving the castle to help his explanation …

Jermaine then pointed out that if you were playing badly you could make the other player win instead, even from C6. From then on all theories had to include, 'If both players play their best moves'. Having proved that C6 was a good place to be, I asked if other squares were important in some way ...

After much staring at the board, 'B7 is the same as C6', said Nicholas. They suddenly noticed the diagonal line emerging, and excitedly tried the castle from the next point along the diagonal, then the next point and so on.

'You win from any square on the diagonal', they eventually decided. I asked whether they wanted to continue their original test of whether the player who placed first always won. They said they did and discovered within seconds that wherever the piece was placed, as long as it wasn't on the diagonal square H1, the next player can always move it straight to a square on the diagonal and therefore win.

'Would you like to carry on playing?' I asked. 'There's no point,' they said, 'It'll be boring.' 'Could you make the game more interesting?' I tried. After an exchange of ideas they wrote:

Some ideas for new rules

Move one square at a time
Move two squares at a time
You could start only on white
You could change the piece to bishop, king, queen, knight
You don't have to move nearer the winning square

Shirley Clarke added that, 'This kind of maths is so exciting for children. As Jermaine was writing their 'ideas for new rules', Nicholas was compulsively trying out his own ideas', and concluded,

(this) theory about games seems to be true! The children started off by experimenting to test their theories but, in playing the game, it gradually dawned on them that there were mathematical reasons behind what was happening; they had moved on to the idea of proof.

(Clarke & Wells 1988:4–5)

Sammy Freeman and the magic squares

2	7	6
9	5	1
4	3	8

The three by three magic square

The familiar puzzle on the left is to arrange the digits 1 to 9 in a three by three square so that all rows and columns and both diagonals have the same sum. It allows pupils to argue logically as well as attempt solutions by trial and error. With a little reasoning, pupils can prove that,

- The sum of the rows is 15. This is *relatively* easy, since the sum of all three rows is 45.
- That the middle number must be 5. This is much, much harder. The four lines through the centre sum to 60 and include the central cell four times and all the other cells once.
- That all solutions are 'basically' the same, and that there are eight of them. This is also much harder: what does 'basically the same' mean?

Many pupils with a simple sense of 'rightness' and symmetry will think of putting the middle number, 5, in the middle cell, which makes the solution much easier, and so perfectly illustrates the value of a nascent aesthetic sense.

Sammy Freeman was in his last year at Highbury Quadrant Primary School. He was extremely strong and highly imaginative and so was able to use analogy much more powerfully.

He started by claiming — being familiar with three by three square — that a four by four square was impossible. His reasoning was complicated and I could not understand it so, knowing that he must be mistaken somewhere, I eventually showed him a four by four square and he agreed that his argument must be wrong.

Sammy then immediately announced that he would fill a five by five square by starting, in effect, with a three by three square in the middle. He calculated the sum of 1 to 25 and found that each line would sum to

65 and so the middle number would be 13, so he added 8 to each cell of the three by three to get this starting position:

	10	15	14	
	17	13	9	
	12	11	16	

A strategy for a five by five magic square

He then started talking about the 'scale'. He considered 5 as the middle of the sequence 1–9 with four numbers on either side of it, and compared it with the sequence 9–17 with 13 in the middle and eight numbers on either side of it. He deduced that 18, the lowest of 18–25 should go where 6, the lowest of 6–9, went in the three by three square, i.e. in the top right corner. He then deduced by a similar but more complicated argument that the top left hand corner, should be 20. He then spent about 40 minutes juggling the remaining pairs to produce this solution:

20	3	2	22	18
25	10	15	14	1
7	17	13	9	19
5	12	11	16	21
8	23	24	4	6

A successful analogy, concluded

This is a perfect example of exploitation of analogy by an imaginative pupil which might also have been produced by some older secondary pupils (1986a:14).

A billiards investigation

In the 1930s a group of enthusiastic young mathematicians used to meet at the Scottish coffee house in the Polish city of Lwow where the waiter was entrusted with a notebook in which they recorded original problems. Among the many gems in the book is this one, proposed by Auerbach and Mazur in 1936:

> Suppose that a billiard ball issues at an angle of 45° from a corner of a rectangular table with a rational ratio of sides. After a finite number of reflections from the cushion will it come to (rest in) one of the remaining three corners?

(Mauldin 1981:229)

Hardy and Wright (1960:379) discuss this analogous problem:

> The sides of a square are reflecting mirrors. A ray of light leaves a point inside the square and is reflected repeatedly in the mirrors. What is the nature of its path?

George Polya in Volume 1 of *Mathematics and Plausible Reasoning* gave the related problem of deciding in which direction to strike a ball on the edge of a rectangular billiard table so that it returns after three bounces to its original position. Polya solved the problem by reflection. The Scottish Coffee House problem is now familiar as the billiard ball problem or investigation, and can also be solved by reflection:

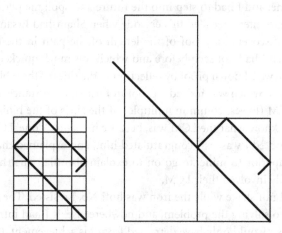

The billiard puzzle, solved by reflection

It appears in popular published collections such as *Starting Points* by Banwell *et al.* (1972) and was discussed by Dietmar Küchmann (1985:2–3) in 'Transforming Billiards into Diagonals' where he showed that the billiards problem is isomorphic to the problem of finding how many cells the diagonal of an integral rectangle passes through, identifying two well-known problems and linking the solution to Polya's problem.

Shamshad Ersan, who was 12 years old at the time, had a different argument which has the virtue of being extremely simple. In the five by

Shamshad Ersan: a moral

'Once upon a time, many years ago — the fairy story format is appropriate — I arrived at school to find a girl waiting for me. I knew her quite well, in so far as anyone knew her. She was gawky, and a loner, and lacking in confidence and achievement. She told me that she had learnt her — I forget which one it was — 'times table'. Of course, I asked her to repeat it to me. She did so, perfectly, with a glow of pleasure on her face. I told her, 'Well done!' and then suggested that she now try to learn the next 'times table' in sequence. In an instant, her face fell, her pleasure evaporated, and I realised that I had blundered very badly.

When she waited for me to arrive, full of expectation, she only wanted me to acknowledge her achievement in mastering one of her tables. She didn't want to be told that there were more tables ahead — she already knew that better than I did, because each table to her was a small mountain to climb, whereas to me it was just an item. But the pleasure of the moment wasn't good enough for me, and I had to step into the future and spoil the performance.

Many years later I recalled this episode when Shamshad Ersan, after eight hours work, discovered a proof of the length of the path in the billiard ball problem which I had not seen before and which was much quicker and neater than the usual well-known proof by reflection in the sides of the table. However, his explanation, which was indeed a proof in form and substance, involved, in effect, the LCM (lowest common multiple) of the sides of the billiard table — but he didn't know what the LCM was, because he hadn't 'done' LCMs.

How tempted I was, as I congratulated him, and explained that his proof was much superior to mine, to go on to explain also that what he had been doing ... really involved their LCM.

Should I not strike while the iron was hot? No, I resisted. The heat was in his reaction to solving the problem, and nowhere else. If I had introduced the idea of LCMs I would only have detracted from his achievement, (which) was actually, for a 12 year old, quite extraordinary.

Pupils should be allowed time and space to enjoy their successes.'

(1992:57)

seven rectangle above, suppose that the ball shoots out of the lower left corner. In order to end up in any other corner it must traverse the width of the table a whole number of times, so the distance it travels is one of, 7, 14, 21, 28, 35, 49 as Shamshad explained the situation. However, it must also cross the rectangle vertically a whole number of times, and so the distance must be one of, 5, 10, 15, 20, 25, 30, 35, 40 …. Therefore the distance travelled is the first common number in these sequences, 35.

At first Shamshad could not say in which corner the ball ends up but when the question was put to him, he could predict that it ends up in (in this case) in the top left corner, because it has traversed the table horizontally an even number of times and vertically an odd number (1986a:14).

'Gary' and the two overlapping maps

'Gary' was a 16-year-old CSE student and self-confessed 'dosser' who was given issue #3 of *The Problem Solver* by his teacher, Barbara Edmonds. He became fascinated and ended up staying up all night, according to his mother, to finish his account of eight of the problems which he described in a 36-page hand-written account including more than 30 figures. Here is Gary's solution of the cover problem in that issue of the overlapping maps.

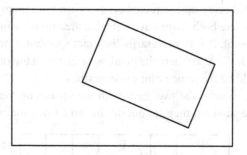

Two similar overlapping maps

This figure shows two maps, one lying on top of the other. Each map shows exactly the same area of country but to a different scale. Is there a point on the map which is exactly above the corresponding point on the bottom map?

Note that his solution was satisfactory to Gary, and should be accepted as such, at his level. It is highly imaginative and very effective, but not perfectly accurate. The next step would be to divide the final pair of squares by even finer grids and so improve the approximation — but that would be going far beyond anything that Gary had ever met, so it not surprising that he missed it.

Gary's work illustrated another feature of motivation: his enthusiasm gave him the time necessary for his achievement.

'Gary's solution to the overlapping maps problem

'I think I will be able to solve this problem but in order to solve it I must first simplify it down into grids or squares. But they must be in the same scale or it won't work.

So first of all I have to find out the measurements of the two maps. The big map is 100 mm by 57 mm. And the small map is 50 mm by 28.5 mm.

This makes the problem easier because the big map is twice the size of the small map.

Now I have to split it down into squares or grids.

This is done by dividing the sides of the map by five and drawing the lines diagonally or horexontally (sic) across the map whichever the case may be [see figure below].

[Gary actually divided both maps into 20 parts on the long sides and 12 parts on the short sides and shaded them like a chess board. We have divided the sides into fewer parts, for clarity.]

I have shaded every other square because when both maps are put on top of each other, where both maps have a shaded area on both maps it could be the same exact spot. It isn't necessarily the exact spot but it narrows the percentage down a little. Also where the blank squares are on top of each other the same point could be the same point underneath.

Now I have to draw the two maps with the squares on top of each other. They will not be shown if they are not on the same row running downwards.

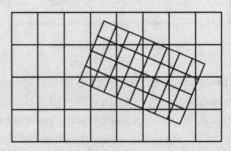

The maps become overlapping grids

(*Continued*)

(Continued)

Where two squares are matching they have been coloured red. [Not in this simplified copy.] From this selection attempt I have narrowed it down into any of nine squares.

To solve the problem and get the one correct answer, all I have to do is to see which of these nine squares is in the same row running across-ways. [Gary then tested each of the nine points and found the only one that fitted his criterion.]

It isn't exactly that point but it is getting pretty near the exact point. Now all I have to do is to find out whereabouts in that square the exact point is [see figure, above]. The point that is circled in the above figure is the exact same point on the top map, and on the bottom map.'

Gary's motivation

Gary the dosser was very clear about his negative attitudes to school, as he explained to Barbara Edmonds:

> I like maths, but I doss because the teachers let us. I could have got a higher grade if the teachers were a bit stricter on us, but if the teachers were stricter I couldn't work for them because strict teachers think they own us,
>
> (Edmonds 1983:5)

perfectly illustrating the difficulties faced by teachers of dissident and unmotivated pupils from what is coming to be called 'the underclass', whose members lack extrinsic motivation, have few prospects and few employment opportunities and little incentive to learn anything at all.

Barbara Edmonds noted: 'Gary disliked school intensely because of its authoritarian regime.' Gary said that he liked maths, but for various reasons he was falling far short of his potential. He continued to fall short after this achievement. With a choice between continuing to study mathematics and joining his mates, he stuck with his mates and gave up maths. A pity? Who can say? In the same article Barbara Edmonds quoted Laurie Buxton from his book, *Do You Panic about Maths?*

> ... there is, in fact, a much higher level of anxiety among students than many teachers would credit, and this is true even when the relationships (between student and teacher) are particularly good.

She later quoted Laurie Buxton's conclusion:

> Mathematics is a largely contemplative and exploratory subject and the
> extent to which it can be an inner private pursuit is not always understood.
> It cannot be pursued without considerable input, but the satisfactions
> should lie in success in pursuing it, not in a teacher telling you it is right.
> *The authority lies in the subject*, and the fundamental aim of the operation
> is that the students should learn, understand and actually *do* maths.

Jennifer Kano's solution

Gary was indeed doing mathematics, and so was Jennifer Kano who met
this same problem while doing a mathematics course at Cape Cod
Community College in Massachussetts that used *Hidden Connections
Double Meanings* (1988f) as a textbook. All of the problems set at the end
of Ch. 11: 'One problem, many solutions' referred to the two overlapping
maps puzzle and asked readers to say which of several arguments for finding the common point were correct.

The first suggestion was Gary's, covering the maps with matching
grids and searching for the common grid squares. The second argument
concerned a one-dimensional case in which two rulers with different
scales were placed next to each other. The third argument concerned maps
which overlapped too little or in the wrong way, and so had no common
point. The fourth suggested drawing a third tiny map with the same relationship to the small map as the small map had to the larger: and then
repeating so that the 'incredible shrinking maps' tend to a limit.

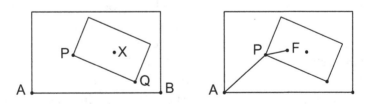

Where is the common point?

The fifth argument was that if X is the common point (left-hand figure)
then triangle AXB is similar to PXQ, and X can be constructed, and finally

the sixth argument was that APF (in the right-hand figure) is the start of a spiral which will tend to the common point as limit.

Jennifer Kano was not satisfied, however:

> I discovered a very nice solution …. after working with the grid system you suggested in number one … When I went to my next class I was rather disappointed to find that my solution was not the 'correct' solution. I reviewed my work and still felt that my solution was viable, so I went to my professor, David MacAdam, and asked him to look at it … He told me that it was possible that my solution was one that no on else had come up with before and suggested that I write to you and ask if you have seen this solution before.

No, I hadn't seen it before, though it is in Coxeter's *Introduction to Geometry* (Coxeter 1969:74). Her proof goes like this:

> My answer is, yes, as long as the areas outside the rectangles are considered to be part of the map there will always be a common point,[§] although that point will not always be inside the rectangles.

> To find the common point, extend corresponding sides of the two maps until they intersect. Then draw lines between the intersections of the opposite pairs of sides. The intersection of the two lines thus formed determines the common point. This always works because the two points where the long sides (extended) of the small map intersect the long sides of the large map determine a straight line along which every point has the same first co-ordinate on the small map as on the large map: and similarly, the two points where the short sides (extended) of the small map intersect the short sides of the long map determine a line along which every point has the same second co-ordinate on the small map as on the large map. The point of intersection of the line of common first co-ordinates and the line of intersection of the lines of common second co-ordinates will be the common point of both maps — the point having both co-ordinates in common.

[§]Which might be at infinite if corresponding sides are parallel.

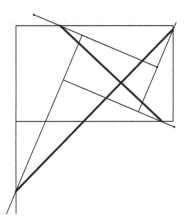

Jennifer Kano's beautiful solution

Notice the imaginative leap in her first paragraph — the qualification, 'as long as the areas outside the rectangles are considered to be part of the map' is also highly imaginative and creative — and the manner in which the pattern of matching grids which was her starting point retains its ghostly presence in her argument.

Note also the *power* of her argument which she also appreciates aesthetically: it is 'very nice'. It certainly is! There are other proofs of the result beyond those sketched here, but I cannot believe that there is another as simple and elegant as Jennifer's.

I replied to Jennifer Kano to point out that her disappointment was unwarranted, that the whole point of the chapter was to illustrate how one problem could have several solutions and that her additional and very beautiful solution therefore reinforced that point!

On deciding to publish her proof in this book I contacted her again: her inspiring letter in reply is reprinted in the Appendix.

Daniel Schultz and the two 'parallel' lines

It is a commonplace that many geometrical theorems have been found by experiment and another that one figure 'proves' nothing because there are many coincidences in geometry, but if you redraw the figure and then redraw it again, you can reasonably conclude that the feature is real — and think about investigating it further, and proving it. The next figure shows that this conclusion may be false. Daniel Schultz, aged 12, first drew a triangle of no particular shape, then divided the sides into quarters and joined the points to form a star. He then noticed that the two bold lines seemed to be parallel.

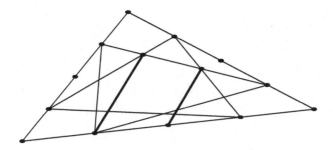

Are the bold lines really parallel?

On redrawing the figure, starting with different triangles, the two lines seemed always to be parallel, but this was an illusion. A calculation showed that they were 'typically' about 1/3 of a degree apart.

In order to explore this phenomenon better, I taught Daniel to use weighted averages to find the midpoints of given lines and indeed the co-ordinates of points dividing a line in any given ratio. Daniel was able to calculate the slopes of the lines and prove that their 'parallelism' was no more than a convincing illusion (Wells & Schultz 2008b).

7

Ways of Seeing Mathematics

Three very general concepts or similes — or metaphors — illuminate the nature of mathematics: it is like an *abstract game*, a *science*, and is a question of *seeing*.[5]

As an abstract game or collection of games, mathematics requires the player to see and observe, to make moves, to plan ahead, to develop an understanding of tactics and strategy and appreciate the power and elegance of mathematical arguments much as a good game of chess can be appreciated. Unlike the chess player, however, mathematicians repeatedly change the rules of the game, add new rules, new pieces and so on.

As a science, mathematics can be explored by observation and experiment, by creating hypotheses and testing them.

Both of these images of mathematics imply that the mathematician must be continually *seeing*, making observations, looking ahead (as chess players would put it), spotting differences and similarities and making analogies.

Learning mathematics and abstract games

There are illuminating comparisons between learning mathematics and learning to play an abstract game like draughts or chess. Learning the rules in relatively easy, but the beginner then discovers that in order to play even

---

[5] This is the subject of *Mathematics and Abstract Games: An Intimate Connection* (2007), Rain Press, since republished by Cambridge University Press in 2012 as *Games and Mathematics: Subtle Connections*.

moderately well you have to play enough to develop some kind of judgement and intuition which will at least vaguely suggest to you what your next move might be.

So it is no surprise that if we observe novice chess players we see the same combination of steep learning curves and steep curve of differences in ability that is observed in mathematics. It is no surprise either, therefore, that while just about anyone can play chess after a fashion (if they wish) a strong attraction towards abstract games is found in only a few.

This is a pointer to the value of *applications*, especially to science, as a motivation for many pupils. The science of the natural world is naturally rich and vivid, and mathematics becomes richer and more vivid when linked to science for pupils whose magnetic attraction to abstraction is weak.

The claim that 'mathematics is a language'

This is just as unreasonable as the claim that physics is a language or biology is a language — claims which are never made. Physics has a special language with many novel terms not found elsewhere, and it also uses mathematics and therefore the language of mathematics — but physics is not a language but an activity which tries to model the world.

Mathematics is an activity which tries to understand mathematical situations, and to solve problems, and then apply the rules to the world. Mathematicians *use* their own language to talk about what they are doing and other people use this a language also, more or less, but if mathematics were nothing but a language it would be effectively useless.

Cockcroft, remarkably, wrote (Cockcroft 1982:1):

> We believe that all these perceptions of the usefulness of mathematics arise from the fact that mathematics provides a means of communication which is powerful, concise and unambiguous ... We believe that it is the fact that mathematics can be used as a powerful means of communication which provides the principal reason for teaching mathematics to all children.

No it doesn't! That would only justify teaching them the language of mathematics — as if a biology course in school might consist of teaching only the language of biology!

There is another obvious link: 'play' is essential for success in mathematics and games. There is no substitute for learners at chess playing the game, and no substitute for pupils playing mathematical games, such as elementary algebra, in order to develop rich personal meanings and intuitions, both cognitive and affective.

Many people find abstract games such as chess and go (and many card games) enjoyable — but many do not, so an understanding of the connection may highlight why some pupils are not motivated to do well, or to try to understand those aspects of mathematics which are indeed most game-like!

Rules and their unexpected consequences

Using rules is not limited to game players or mathematicians. Artists and architects, for example, use rules. Classical architects used them to ensure that his buildings were classically 'correct', ensuring their beauty. The modern constructivist may use rules to decide how to create a work of art, or to arrange its elements, because by using rules the personal element is (partly, not completely) eliminated and the artist finds that following self-imposed rules 'creates' situations that the artist's own imagination would never have considered.

Anyone who tries to invent an abstract game soon experiences this phenomenon: you choose a board, or design a pack of cards, think of some rules and start to 'play the game'. The situations you create with the rules are *potentially* there the moment you decide on the rules — but you only discover them by your own exploration (or by exploring with a friend).

The box below shows 14 similarities and seven differences between mathematics and abstract games (2012:Ch. 3, 5, 6).

Mysterious connections between abstract games and mathematics

- They are abstract and can be analysed in the head.
- They are abstract.
- They are about a novel kind of object.
- You have to 'look ahead' and 'see' mentally.
- There are possibilities of proof.

(Continued)

(Continued)

- They can be simple, elegant and beautiful.
- They can scientifically explored.
- The game pieces are defined solely by the rules.
- The rules have to be consistent.
- Games need tactical and strategical understanding.
- Structure, forced by the rules, makes game playable.
- Players use analogy to generalise and specialise their past experience.
- Players require intuition AND imagination AND formal reasoning.
- Players make mistakes, including mistakes in formal reasoning.
- Competition: if they are competitive, they are so in different ways.
- Asking questions about: mathematicians continually ask questions about their situations and so 'step outside the box'.
- Creating new concepts and new objects: mathematicians do this much more than chess players.
- Increasing abstraction: mathematics becomes more and more abstract the farther you pursue it.
- Finding common structure in different miniature worlds: chess is one miniature world, mathematics contains many, but with much in common.
- Mathematics is used everywhere in the hard sciences: games are not.
- Certainty, truth and error can be found in abstract games and mathematics, but errors in chess and go are simpler.

Mathematics as game-like: the game as a miniature world

> It is difficult to give an idea of the vast scope of modern mathematics ... I have in mind an expense swarming with beautiful details, not the uniform expanse of a bare plain, but a region of beautiful country, first seen from a distance, but worthy of being surveyed from one end to the other, and studied even in its smallest details, its valleys, streams, woods and flowers.
>
> (Cayley 1883/1896:449)

Mathematicians can be seen as working within miniature worlds. Each has its own rules, and can be developed and expanded by creating and adding to it new ideas and concepts, new objects and new categories. Mathematicians are continually *asking questions about* their miniature worlds, changing them, adding new features and finding relationships between different worlds. (That is one striking difference from chess in which the rules, if you are playing a standard variant such as the occidental chess recognised by FIDE, change extraordinarily rarely and only, in recent decades, in minute respects.)

These miniature worlds can be explored in a scientific manner, seeking and collecting data, forming hypotheses and testing them. The result ought to be that pupils become familiar with the landscape. The headmaster of a secondary school was once asked (a long time ago) what he expected of the pupils who were coming up from his primary feeder schools. He replied that he would like them to be friends with the numbers 1 to 100. The Cockcroft Report had a similar conclusion: 'numerate' should imply 'at-homeness' with numbers (Cockcroft 1982:11).

However, mathematics differs from the natural sciences because it has recourse to proofs which not only create extremely high levels of certainty, far beyond those available to the natural scientist, but also by depending for their creation on new ideas and concepts, 'force' the mathematician to understand the subject more and more deeply. The best proofs, as we have seen, not only create confidence but provide illumination.

Finally, abstract games can be very beautiful though aesthetic and affective judgements in games vary from person to person. There is simultaneously a high degree of agreement, and a very wide range of disagreement suggesting that their own preferences will have a special value for them.

Games, tactics and strategy

To play the games of mathematics well, pupils need to explore their miniature worlds in order to develop their own repertoire of tactical and strategical concepts without which they will be lost. These tactical and strategical ideas will inform the pupil's judgement and mould their expectations. This development takes place both through the player's own experience, and vicariously through the experience of others. It is a scientific process, in which hypotheses are formed about different types of positions, tested against experience and modified or abandoned. (The same approach is more or less taken for granted by players of sports, such as tennis, football, or mountaineering.)

Unfortunately, the very idea that tactical and strategical ideas exist in mathematics or that strategy can guide tactics or that tactical analysis will determine whether strategical ideas were good or bad 'theories of the position', is absent from mathematics teaching.

The metaphor of 'tactics and strategy' is also a suggestive guide to some of the difficulties that pupils, and therefore teachers, may face in tackling mathematics. How do pupils develop strategical/tactical ideas in mathematics? How can this development be aided? What tactical ability can be expected of pupils? Such questions are very different from the problems of getting traditional pupils to reproduce textbook content, or even to do maths scientifically.

Abstract games and game-like mathematics require the ability to look ahead, to foresee the results of the moves you are planning. This ability is in weaker players poorly developed. When I taught chess in evening class, several enthusiastic players could not see more than two or three moves ahead, or recall what had taken place two or three moves earlier. They saw every position in a 'fuzzy present'. Naturally, being unable to perceive a game as a whole they could not follow strategical ideas developing over many moves.

The analogy to physical games and sports has other implications. Pupils who may have never thought that mathematics needs insight and imagination, or intuition, can appreciate that they need these qualities in the games familiar to them. They can also appreciate the idea of a brilliant move, or even a beautiful move. They are likely to agree that a shot at tennis can be beautiful and this appreciation can take them part of the way to appreciating aesthetic values in mathematics.

Elementary algebra as a game

Pupils who are introduced to algebra-as-a-game appreciate that they have a choice of moves, that some moves are better than others, that they need to look ahead, that look-ahead depends on their past experience, and that practice is required, not the skill-practice which hones a skill to perfection, but game-practice which develops their understanding of the miniature game-world.

As we have emphasised, playing the game of algebra is difficult for most pupils. A few strong pupils will take to it like a duck to water and at once swim confidently and take for granted that — for example — you have a choice of moves at each step, and of course you actually choose the move in the standard algorithm (for example) because it gets quickly to the solution and not because it is the only move available.

But while the stronger pupils takes such ideas for granted, weaker pupils do not possess them at all and, their experience also being limited, they are unlikely to develop them by themselves. Meanwhile they are reduced to trying to remember how the teacher did it, and often actually trying to read the teachers mind as a substitute for using their own mind and their own understanding, which is too poorly developed.

Algebraic tactics and strategies

Once you understand the basic rules of algebra you have a choice of moves to play. Find a good sequence of moves and you may solve the problem. In this position from the 'game of elementary algebra' there are many possible moves:

$$2x + y = 40$$
$$x + 3y = 30$$

It follows that,

$$4x + 2y = 80$$
$$6x + 3y = 120^*$$
$$\cdots\cdots = \cdots$$
$$2x + 6y = 60^*$$
$$3x + 9y = 90$$
$$\cdots\cdots = \cdots$$
$$x - 2y = 10$$

and also that,

$$3x - y = 50 \qquad \text{and so on} \dots$$

Unfortunately, pupils instructed to follow the *standard strategy*, are almost always directed at once to one of the starred equations and so they fail to appreciate that there are in fact a large variety of possible moves. If the objective is to solve the equations as fast as possible, this does not matter — but if the objective is the wider one of helping pupils to understand the role of algebraic manipulation then it is a grave error.

The next equations seem more complicated, having three unknowns, but if you *see, notice* or *observe* the symmetry, then the solution is easy:

$$a + 2b + 3c = 10$$
$$3a + b + 2c = 18$$
$$2a + 3b + c = 14$$
$$6(a + b + c) = 42$$
$$a + b + c = 7$$

and so $b + 2c = 3$ and $3a = 15$ etc.

This exploitation of symmetry is not a 'cheap trick' but rather a basic tactical recourse, or even a strategical goal, since other, deeper 'tricks' are the basis for powerful general methods of solving all such equations.

Elementary algebra plus the more advanced topics of calculus and matrices and vectors, have a pre-eminent role in applications of maths to science so it is fortunate that large classes of algebraic problems can be solved by well-understood techniques.

Euclidean geometry is also much like a game but it is not used to the same extent in applications (though vector algebra models some of its features) and it cannot be reduced to the use of powerful techniques. It does brilliantly illustrate, however, the other aspects of mathematics as a collection of game-like miniature worlds. In particular, it contains many fascinating objects (far more than *elementary* algebra) with surprising properties that pupils can explore scientifically as well as investigate by reasoning, and which can often be proved, less by fluent technique than by imaginative arguments.

The need for interpretation

The chess or go player is continually stepping back and interpreting the position, trying to work out what it *means*. Algebraic arguments are very powerful but the results also need interpretation. Here is an example:

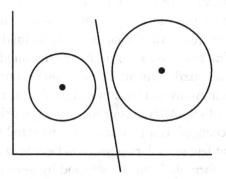

What is the meaning of the straight line?

The equations of these two circles are,

$$(x - 3)^2 + (y - 4)^2 = 4 \quad \text{and} \quad (x - 9)^2 + (y - 5)^2 = 9$$

If the circles intersected, then we could find and draw the equation of line through the two points of intersection by subtracting the equations. Let's do it anyway, to get the equation,

$$6x + y - 38 = 0$$

This real line has been drawn on the figure. What does it *mean*? Since we have only acted as if the circles intersected, this must be a matter of interpretation: sticking to the real plane, one interpretation is that the common line is the locus of the centres of all the circles which intersect both the given circles at right angles. If we allow complex numbers, then this real line does indeed pass through the finite complex points of intersection of these two circles (1995b:133).

What's the point of quadratic equations?

We have already suggested that quadratic equations are one of the peaks of difficulty and abstraction for most pupils, while also being totally unmotivated. What do they mean to most pupils? They are an extremely rich source of ideas and problems — but not the way they are presented in textbooks.

Why are they on the syllabus? Why do pupils need to solve them? They are not told. Is it because just a few pupils will need them in years to come? Are the interests of the majority being sacrificed once again to the minority?

The solution of quadratics by factorisation is fundamental from an advanced viewpoint, but to most pupils it is a trick and is only learnt as a rule with a very small puzzle element. The solution by completing the square also appears as an unmotivated trick though to the expert it is equivalent to shifting the axis of symmetry to coincide with the y-axis, x = 0. The roots are essentially unchanged, can be calculated easily and then the original roots recovered but few secondary pupils will be able to appreciate this.

The quadratic formula is not understood by most pupils but is also memorised as a rule — another pointless exercise.

Quadratics as presented in standard textbooks or as demanded by school examinations are not puzzles because too much understanding of notation is required before they tackle them. All puzzles should be approachable immediately. We should also bear in mind the Cockcroft claim that the syllabus should be built up for the benefit of the weaker pupils not watered down from the top. The least we can do, therefore, as long as quadratics are imposed on teachers and their pupils via the syllabus is to present in as motivating a manner as possible. The two obvious approaches are via puzzles and via science (which we will look at in Ch. 11).

Quadratics as puzzles

Pupils can solve quadratics-as-puzzles themselves, make observations, develop their experience and so encounter important ideas. They can exploit their experience to solve quadratics by inference and insightful trial and improvement: to represent the values of quadratic expressions visually and explore their shapes. 'Quadratic equation' puzzles can be introduced in integer terms to primary age children who understand the square of an integer.

A puzzle: I think of a number. The square of the number plus 10 equals 7 times the number. What number did I think of?

A powerful metaphor for introducing algebraic notions is the idea of a code. Writing this puzzle in algebraic *code* using N for the number, it becomes,

$$N^2 + 10 = 7N$$

This is relatively easy to think about because the code is so simple. This format is also much easier for most pupils than the more abstract form, $x^2 - 7x + 10 = 0$, let alone the completely abstract and general form, $x^2 + bx + c = 0$ or the perfectly general form, $ax^2 + bx + c = 0$.

Note that this is a self-contained puzzle — and not a means of 'teaching' something more sophisticated. At the same time, it is an example of *surreptitious* learning. *Some pupils* may possibly, *not necessarily*, take something extra away from solving this puzzle.

By trial and improvement the answer could be 2, but I could also have thought of 5. After trying several such puzzles pupils may notice, without being able to explain why, that $2 + 5 = 7$ and $2 \times 5 = 10$, the two numbers in the original puzzle.

This discovery in itself introduces two important general concepts, that some problems have more than one solution, and that these varied solutions may be connected. In this case, if you know the connection, and one of the solutions, you can at once find the second solution. The observation also leads to an easier method of solution — just look at the factors of 10 which add up to 7.

There are many other puzzles about quadratics.

More questions about quadratics

How many solutions does this equation have? $N^2 + 15 = 8N$
How many solutions does this equation have? $N^2 = 2N + 15$

If $N^2 + 10 = 7N$ find a hidden connection between the solutions and the numbers 10 and 7.

How many solutions does this equation have? $N^2 + 11 = 7N$

Find a quadratic equation with only one solution.

Find a quadratic equation whose two solutions are 4 and 5.

If a quadratic has integral solutions, they must divide the constant term. Why?

If the graph of a quadratic equation just touches the axis, how many solutions does it have?

(Continued)

(Continued)

Which quadratic equations *always* have at least one solution?

Which quadratic equations have two solutions, very close together?

A quadratic expression such as $x^2 - 10x + 18$ does not appear to be symmetrical but its graph is symmetrical. Why?

The quadratic $x^2 + 5x + 6$ can be regarded as the curve $x^2 + 6$ *sheared* parallel to the y-axis. Why does shearing not change its shape?

Can you find a quadratic equation with solutions $4 - \sqrt{5}$ and $4 + \sqrt{5}$?

How many different shapes can a quadratic graph have?

Plot a cubic equation. What kind of symmetry does it have, if any?

Illustrations of the solution of a quadratic puzzle

All these puzzles can be posed to pupils who have not yet learnt about negative numbers. Indeed, some of these questions could be used to motivate the introduction of negative numbers.

The values of $N^2 + 10$ and $7N$ can also be shown in a table or on a graph:

$N^2 + 10$	$7N$
11	7
14	14
19	21
26	28
35	35
46	42
59	49

There are two matching points

The graph suggests strongly that non-integer values of x might fit the same visual pattern, starting with half-integer trials. They do, of course, but this should be a motivating discovery by pupils themselves: ideally they should be surreptitiously induced to predict that non-integer values of x and y will fit the visual pattern and if they do not find the result *surprising* this ought to be only because their nascent intuition told (some of) them that of course the pattern would continue.

Another way to draw the visual graph is to show the differences between the two columns in the table, or the values of $y = x^2 + 10 - 7x$.

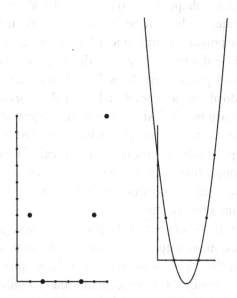

The discrete points have been 'filled in'

The points of this picture can also be 'filled in' by using fractional values of x, but it should be up to the pupils to decide whether they want to draw a continuous line between them, and if they do want to, to explain why all the points on the curve fit the equation of the puzzle (by using their calculators).

This picture strongly invites another question and a natural response: many children will think that it *makes sense* to say that 19 − 21 is two less than nothing — and this is a perfect example of pupils learning by *making sense* of their situation. (It will also fit any previous experience of negative temperatures — almost the only everyday, scientific, context in which pupils naturally meet negative values.)

This is the easiest introduction to negative numbers, after everyday cold temperatures: negative numbers as markers with no calculation on negative numbers themselves.

If pupils do decide that they can 'fill the gaps' to create a continuous curve — or even if they do not — they may wonder whether the pattern could continue to the left.

If pupils experiment by making up their own analogous puzzles they will observe that all the shapes seem to be symmetrical, up to the brick wall on the left-hand side. Could there be anything beyond the brick wall? This puzzling question potentially introduces further ideas of negative numbers by a second (and harder) route. By drawing the line of symmetry, pupils can construct points 'past the wall' and so continue the parabolic shape, but how do these new points relate to their original table? This raises, of course, hard puzzles. At this point the subject of negative numbers can be left, *and should be left*. Pupils have been led to negative numbers almost surreptitiously, in a meaningful context. They can now be left for a while — a long while — during which time they ought to 'just happen' to meet them again, surreptitiously, before they get around (if they ever do) to studying them further.

(In an ideal world — or even in the present world organised on more rational principles, many pupils would *not* study them further because they will never use them and the ideas met here (examples of general concepts) are much more valuable to them than some *technique* of calculation with negative numbers which they will never require.)

In the meantime, there are many more thought provoking questions about quadratics, both simple and complex. Finally, among the hardest problems for almost all pupils are those concerning the use of factorisation, completing the square and other 'hard' methods of finding solutions to an arbitrary quadratic — which few pupils can understand deeply but will also never need, and do not need to know.

Technique and creativity

WARNING

Almost all mathematical techniques and methods are much too powerful for immediate use in the classroom.

Regrettably, technique in mathematics education has achieved a bad reputation over the years among many teachers. It is commonly thought of as dry-as-dust, as 'mere technique' and assumed to be the opposite of 'creativity'.

General concepts about technique in mathematics

The best game-like sequences of moves can only be found by using imagination.

Powerful technique, like powerful play at chess, depends on experience.

Algorithms tell you exactly what to do, step by step, to get the correct answer (or answers).

Algorithms can be used without any understanding at all of why they work.

If you rely too much on algorithms then you may never develop your own understanding and in the long run, you lose out.

Mathematical *methods* tell you roughly what you have to do — but what you actually do depends on the problem and your own decisions.

If you try to use methods without understanding why they work, you are liable to go astray quickly.

Game-like technique can be improved with the help of a coach, just like technique in any other game or sport.

To solve novel problems, technique is seldom sufficient: you need new ideas, and maybe technique as well.

Equally representative is the view that, 'The traditional, abstract, and formal methods of teaching algebra have not had much success.' Indeed they have not, and this suggests the reason why game-like technique and game-like aspects of mathematics in general have, absurdly, fallen into such disrepute. It is not that technique — correctly understood as a combination of algorithms, methods and game-like sequences of moves motivated by tactical and strategical understanding — is unimportant, the very opposite is the case — but rather that it has proved difficult to learn and apply successfully.

Two reasons seem clear enough: making the analogy with abstract games we must suspect that mathematical technique is indeed difficult to master just as abstract games are difficult to learn and play well: and secondly, what techniques might be learnt in school have almost no motivating applications because of the absence of any connection with hard science — where mathematics has historically been applied most often and most successfully.

As a result, game-like technique has been downgraded in status, which means in turn that less attention is paid to it by researchers and those keen to improve mathematics teaching, so the task of getting children to understand, say, basic algebra, have nowhere near been solved, and so the vicious circle continues.

It is not true that technique is old-fashioned or dry-as-dust. Quite the opposite: techniques represent the quintessence of the mathematics of years gone by, summarised and condensed into ideas and methods that can be applied by anyone who has studied them, without having to reproduce the highly imaginative thinking that created them in the first place.

Formal methods have another feature: they apply to wide classes of problems. They are therefore appropriate for that stage in the pupil's development when he or she has realised that many apparently distinct problems can be seen as 'basically' the same. At that stage, not before, the pupil is ready to appreciate that there is a formal method which will apply to any one of an entire class of problems, without the need to interpret each individual problem.

Of course, it goes without saying that mathematical techniques are usually *taught* in a dry-as-dust manner, from syllabuses and textbooks which ignore their origins, provide no sound motivation and omit their most interesting applications, while presenting them at a high level of abstraction which is beyond most pupils.

(So it is deeply ironic that so many pupils, after years of study, have such a poor understanding of the syllabus that they answer many exam questions by using what bits of technique they have picked up, in the form of 'rules': Do this! Then do that! Now do the other! I have noted already the tendency of many textbooks to encourage pupils to do this by actually offering them the rules that invite the pupil to believe that they can dispense with any deeper understanding.)

When pupils find that they can actually use techniques effectively — and potentially creatively — to solve problems and explore situations, then technique itself becomes motivating.

At the same time, because no distinction is recognised between strict algorithms and — more or less creative — mathematical *methods*, teachers conceive no reason to emphasise technique as an aspect of pupils being creative and being young researchers.

In emphasising the role of game-like mathematics, the teacher once again emphasises the links between the learning of mathematics and the real world of the child-embedded-in-the-world. This includes the need to have at least some degree of mastery of the technique before you can use it — although it is quite false to suppose that technique must be perfected before it is useful — and the presence of strong affective and aesthetic values.

The idea that creativity and technique are sharply separated, as if by a deep chasm or high fence, is entirely false. There is a *continuum* from the highest levels of creativity all the way to the most routine and rote applications of algorithms.

Not only is the assumption of an impenetrable divide between the aspects misleading and dangerous, but pupils benefit greatly by learning that there is a slide, a gradual movement, from one to the other.

From creativity to technique

We can think of the solutions to different problems lying on a scale from the purest creativity to the purest technique. Historically, this is how many solution methods have gone: first understood imperfectly then (as understanding deepened) hesitant solutions were organised and turned into methods and finally — sometimes — into algorithms.

Pedagogically, this is how pupils should approach such problems: first as novel puzzles to be thought about and puzzled over; then as problems which they recognise as being of a type, so that they can exploit their previous experience to solve them more readily but still without knowing about a method; and finally — and not sooner — as problems of standard types with standard methods of solution.

As examples of pure creativity we might take Shamshad Ersan's solution to the billiards problem (page 142) or Jennifer Dunham's solution to the fill-the-cells puzzle (pages 133–4). No methods were used and no rehearsed techniques: the solutions depended entirely on the solvers' original ideas. For the effective use of a method consider weighted averages (pages 148–9). The method is standard and lets the solver perform a number of operations — but the choice is entirely up to the solver. For examples of algorithms, the standard arithmetical sums are perfect but there are many more in elementary mathematics.

Technique and creativity cannot be separated. Technique should not be seen as opposed to creativity, let alone to understanding, but as an aspect of both. Technique helps you to be creative in different ways — technique is like a powerful tool, which enables you to do things you couldn't do before. (So do calculators and computers which is why they are used to take the place of some old and obsolete techniques.)

Hans Freudenthal once claimed that, 'when calculating starts, thinking finishes' (Mason & Sutherland 2002:18). This is entirely false. In many situations, thinking and calculation-that-needs-thinking because it is very far from being automatic, go on together side by side.

Much calculation is game-like and not algorithmic (= automatic) and it should be kept game-like for as long as possible. The error teachers make — following textbooks and under intense pressure to get pupils through examinations — is to move too quickly from insightful solutions to the rules to which they can be 'reduced' and then to leave pupils with the rule, only. Rules are worst in effects on weaker pupils. We can represent the effects of rules like this:

	weak pupils	strong pupils
On the surface	rules	rules
Below the surface	nothing	understanding

Stronger pupils can use rules safely because they have, just below the surface, a firm understanding of why they work, ready to be called upon as necessary, so strong pupils save time and lose nothing by using rules. Weak pupils have little by way of understanding behind their surface grasp of rules, and so they have no foundation of understanding on which to progress further, and when the rule does not precisely fit the situation, they will be lost — or they will charge ahead and use the rule anyway, and be wrong.

To any teacher or educator taken in by the currently fashionable extreme fallibilistic point of view and who therefore fails to appreciate the difference in subjective certainty between game-like and science-like mathematics, technique could have unpleasant negative connotations and therefore when it appears in the process of problem solving, they may be effectively blind to its presence.

This mistake is not made in the field of abstract games or sports, or the arts. No one thinks that technique in football is limited to repeatedly bouncing a ball on your knee. That is one small aspect only. Creative technique in football consists of, for example, trapping a ball, turning and shooting it into the top of the net, in a fraction of a second.

Technique in violin playing does not consist of just playing scales perfectly — though that is as important as the mathematician's ability to make basic algebraic transformations again and again without error — but in using a broader technique to play complex pieces of music.

In the application of mathematics, the depth, insight and intuition of the applied mathematician, whether a physicist, chemist, biologist, engineer, operations researcher, or computer scientist is focused on the application itself.

It would be very surprising to find that an engineer using Laplace transforms to solve differential equations had a deep and intuitive understanding of the pure mathematics behind them, just as it would be unusual to find a pure mathematician who had a strong intuitive grasp of the engineering situation.

8

Euclidean Geometry as Game-Like

> It is the glory of geometry that from so few principles, fetched from
> without, it is able to accomplish so much.
>
> (Newton, Preface to *Principia*)

For nearly 2,000 years, Euclid and his *Elements* of geometry were seen as
the perfection of mathematics and mathematical reasoning. The deep
method by which he presented his conclusions, starting from a few simple and apparently obvious (with one exception) assumptions, seemed to
be supremely logical. The wealth and variety of his conclusions from the
very simplest to the extremely complex — and the far greater number
discovered by his successors — was astonishing, as was their beauty and
elegance.

Euclidean geometry was indeed — or so it seemed — an exquisite and
beautiful miniature world constructed on purely logical principles but we
now know better: the weakness in Euclid's system which he and his successors repeatedly puzzled over was actually a deep flaw in his logic. The
parallel axiom could be replaced by very different axioms to create a variety of other geometries, so Euclidean geometry was not only not 'perfect'
but it was only one of many.

We also know that it had other weaknesses. In particular, by using
visual diagrams, and sometimes several diagrams for one proposition, the
Greeks got round the difficulty that they never explained what it meant for
a point to be between two other points: their diagrams just showed it.

171

All these weaknesses have been repaired in the last two centuries, and yet, and yet, the 'new' logically perfect Euclidean geometry has exactly the same theorems as the original!

To the amateur — and that includes school pupils — the landscape of this extraordinary miniature world is unchanged, though we can now recognise, bearing in mind Sylvester's strictures (page 188) that we have two approaches to exploring it, the scientific and experimental and the logical and deductive.

The miniature world of Euclidean geometry is such a marvellous medium for experiment because it is so rich and visual, and because it illustrates all the features and factors of mathematical 'beauty'.

At the same time, so many inferences, as Newton noted, can be made from so few premises. No wonder Euclidean geometry is also thought by so many to be extremely beautiful.

General concepts about Euclidean geometry

Geometry can be used for practical purposes — to make constructions — but also just to discover and prove properties that have no practical use.

Experiment in Euclidean geometry is often easy, because the relations you find are so convincing.

Proofs in Euclidean geometry are often hard, because they depend so much on perception, and you have to think of original ideas

What you can prove depends on what you assume. You have to start by assuming something!

All proofs begin with assumptions, though sometimes these are taken for granted and not explained.

It often helps to think of the basic patterns of the plane, although most of the time you can't see them.

Euclidean geometry is like a game — you know the moves you are allowed to make, and you choose the best sequence.

There are tactics and strategies in Euclidean geometry.

You can create the whole of Euclidean geometry from a few basic assumptions.

Euclidean assumptions, or the rules of the game

Euclidean geometry has the advantage that the basic assumptions as pre-sented in textbooks may seem obvious — but there are other equally 'obvi-ous' properties that pupils are likely to pick out, if they have enough experience and are given a choice. These are unlike Euclid's, let alone those from some modern set of basic axioms.

The basic properties of parallelograms, for example, are liable to seem quite obvious to pupils after a little experimentation, especially if they have drawn them on ordinary graph paper. They are therefore a suitable choice for the pupil's own assumptions, and it is only much later that they may come to the idea even these basic properties can be proved from yet simpler starting points. In the meantime, making such assumptions may raise the question, 'Are there facts in mathematics which are so simple and certain that you don't need to prove them?'

What might pupils assume?

Traditional textbooks such as *Elementary Geometry: Practical and Theoretical* by Godfrey and Siddons started with a short course of experi-mental geometry, so designed that the pupil would be 'led to discover many geometrical truths which are proved later' and went on to prove a selection of theorems on the basis of a limited number of assumptions which were never explicitly listed. Some of them were simply stated in passing, such as, 'We shall assume that all right angles are equal' (*Ibid.*:64). They were attempting, with some success, to avoid premature abstraction and to focus on the practical.

In contrast, modern treatments such as *Basic Geometry* by Birkhoff and Beatley are extremely abstract, and modern American high school texts which are influenced by them are also absurdly abstract and fail, notoriously, to teach students more than a smattering of geometry — and in particular largely fail to teach them how to *prove* theorems.

What should be done? First, we should accept that since abstraction is indigestible by young pupils it should be avoided as far as possible and introduced only very slowly.

Second, since professional mathematicians do not spend all their time thinking about their basic assumptions neither should pupils. It is worth bearing in mind that most published mathematical papers present mathematics in terms of assumed results, not assumed axioms, the former being far above the foundational level of the axioms. Of course, mathematicians can if necessary reword their proofs in order to show all the most basic assumptions they are making, but this is not normally thought necessary and mathematicians will not thank you for insisting that they do so.

Thirdly, therefore, we should choose basic assumptions that are as intuitively plain and plausible as possible.

A basic grid

One realistic assumption that fits pupils with experience of using graph paper is that you can fit squares together to make a square tessellation, and indeed you can similarly fit scalene triangles or copies of any chosen parallelogram to make a plane-covering tessellation. Here 'plane' is not defined but is distinguished from a sphere, for example. Pupils can see by experiment that this does not work for a sphere. (A sphere can be dissected into eight identical eighths and any number of identical 'lemon slices'. How else can it be cut into identical tiles? Not easily, and they cannot be squares with four equal sides and four equal right angles.)

It is a way of saying indeed, what a 'plane' is: a 'plane' is any surface on which this assumption is true.

In such tessellations, pupils will tend to take for granted whatever appears to be totally obvious, and this probably includes 'vertically opposite angles' being equal. They are also likely to think it obvious that corresponding angles, alternate angles and complementary angles on parallel lines are equal. Why not? At secondary level, that is an excellent basis of reasonable assumptions on which to make inferences, for example, about the sum of the angles of a triangle which they are *not* likely to find blindingly obvious.

What about similar triangles? A need for discussion! But pupils can still draw their own conclusions from their experience, especially of drawing similar triangles on graph paper, and then *assume* their conclusions — which after all fit their experience.

The ASA case of congruent triangles is then just an example of similar triangles in which the sides of the triangle has been fixed 'from the base' and the SAS and SSS cases appear convincingly as a result of attempts to construct triangles with given dimensions.

Don't prove the obvious

In this way we avoid the difficulty of proving the obvious and we also maximise the pupils' motivation to use the facts which they know and grasp intuitively because they genuinely seem to be natural and obvious, to prove less than obvious facts such as the angle sum of a triangle.

To insist that the most basic properties — as seen by a professional mathematician — are proved by pupils is demotivating but to discuss which properties are 'really basic' and can safely be assumed is the opposite — a rich puzzle and highly motivating, as well as leading on, if pupils go far enough, to such deep questions as:

- What are the minimum assumptions that can be made to guarantee the proof you want?
- Can you prove that it is always possible to construct a tessellation out of (say) identical copies of a given scalene triangle?
- What are the fewest assumptions which allow you to prove everything you want to prove?

In this way, pupils are led to raise the very questions that point to the need for deeper understanding.

Tessellations and dissections: the basic structure of the plane

Tessellations and dissections are two sides of the same coin. They are, of course, extremely motivating: sets of tiles have long been produced for children to play with, and even with such simple equipment, it is possible to draw conclusions about geometrical figures (see box below).

They also allow children to construct the simplest tessellations, of squares or repeated triangles which do seem 'as if' they would go on for ever, though the actual supply of tiles is limited.

Computer packages such as Cindarella or Geometer's SketchPad produce such square or triangular grids on the plane at the touch of key, which however bypass the same question — whether such grids can actually be drawn *accurately* to extend *for ever*.

We know that the answer is 'Yes' because we see the Euclidean plane as effectively identical to the Cartesian plane which does indeed consists of a square grid, with subdivisions.

We might say that this square grid (and the related triangular grids) display the hidden underlying structure of the Euclidean plane. So it points to more general ideas of hidden underlying structure long before pupils come to specific abstract examples such as groups and matrices.

Proof by tessellation

Since square and triangular grids effectively represent the structure of the plane, we should not be surprised that many geometrical theorems can be proved by seeing the figure as a part of a tessellation. Often this is a way of seeing that the figure is symmetrical in a significant manner as the Vecten figure (pages 84 and 85) illustrated.

A basic observation is that large triangles drawn on the tessellation can be dissected into smaller identical copies.

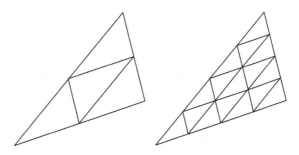

Dissecting a triangle into identical small copies

Such experiments surreptitiously teach concepts of vectors that are themselves, at this level, no more than a way of thinking about and experiencing intuitively the basic properties of the plane.

The simplest such example is the fact that if copies of a scalene triangle tile the plane in parallel strips then the angle sum of a triangle is the angle on a straight line or two right angles. Having proved this property of general triangles, pupils can go use the property without necessarily considering the underlying grid.

A set of children's tiles

Sets of these tiles of three different shapes have been sold as a child's toy — I played with a wooden set as a small boy — and as a puzzle, with designs printed on the tiles which match as the tiles are fitted together.

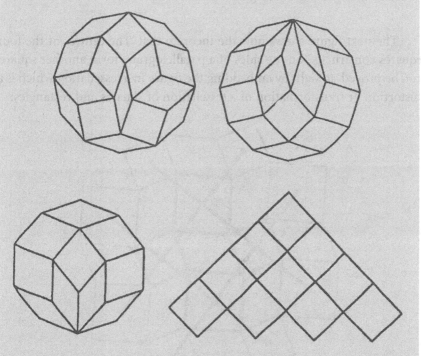

Tilings of decagons and dodecagons

There are many ways to fit the 15 pieces into a regular dodecagon but the top right figure might seem the most natural since it is composed of two strips of rhombuses surrounding a single rhombus.

(Continued)

> *(Continued)*
>
> Looking at the middle six tiles, these form a regular octagon which has been 'squashed'. This suggests a method for assembling 4n tiles into a regular polygon: take the 4(n — 1) case, squash it, and add a new strip of tiles round the outside. This method also works for 4n + 2 edges.
>
> The totals of 10 and 15 tiles hints at the triangular numbers and we indeed can total the number of tiles, 1 + 2 + 3 + ... by counting the tiles in each strip, or realise that the pattern can be hinged at its vertices and open out the lower left decagon as on the right, by breaking the bottom vertex, making the 'triangular numbers' pattern quite obvious.
>
> We can also use the dissection of two or more dodecagons to create a larger one, for example, four dodecagons dissected into one. (1988c:22–23.)

The next figure shows how the theorem that 'The centres of the four squares constructed on the sides of a parallelogram, form another square' can be proved-at-sight by embedding the figure in a tessellation which is a distortion or transformation of a tessellation of squares and rectangles:

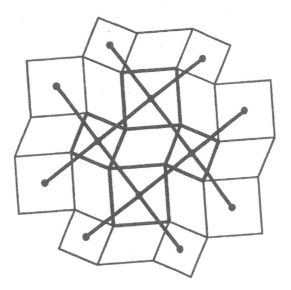

The symmetry of the tessellation is revealed

This tessellation unlike the original figure has the centre of each square as a centre of four-fold rotational symmetry.

A 'continuity' argument

Proving theorems only for figures with their vertices on grid points might seem a very limiting constraint, but this is not so. If we draw a figure and then impose a grid on it, then by making the grid fine enough we can ensure that the vertices of the figure are very close indeed to a grid point: in fact by making the grid finer and finer we can 'make sure' that they are close as we choose.

At that point the pupil is likely to be satisfied, which we should accept because the argument is indeed sound, though a professional mathematician will have a lot of detail to add. We might say that when pupils accept such arguments they are showing sound intuition.[‡]

Proof of triangle properties by tessellation

A triangle has many 'centres': here is the simplest:

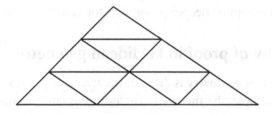

A dissected triangle reveals one centre

We dissect the original triangle into an odd number of identical small copies, nine is the simplest, and the 'centre of symmetry' of this dissected triangle is also where lines joining the vertices to the midpoints meet. 'Centre of symmetry' is in quotes because this 'centre' is not a centre of rotational symmetry at all, unless the triangle is equilateral: yet there clearly is symmetry in the figure, and in some sense it is 'rotational'.

If the triangle is indeed equilateral then a similar dissection shows that the sum of distances of any point from the sides of an equilateral triangle

---

[‡] Compare Gary's solution to the overlapping maps problem, pages 143–5.

is constant:

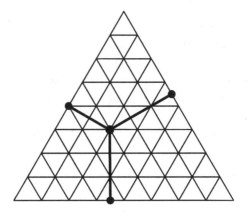

A constant sum, proved by dissection

With each 'step' you take along the edges of the grid, the sum of the three distance is unchanged. As you move from one point to another, one of the lengths increases, another decreases by the same amount, and the other stays the same. This is a much deeper proof than the common proof by areas.

The difficulty of proof in Euclidean geometry

Proof in Euclidean geometry is difficult because it requires so much imagination. For example, the theorem, 'Angles in the same segment are equal' is motivating because it is so surprising and only needs the construction of three lines to create the isosceles triangles on which the standard proofs depend — and yet to 'see' this construction is difficult for many pupils.

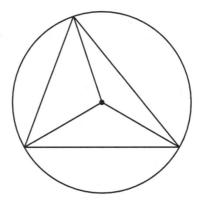

The 'angle in the same segment' theorem

One reaction is to go back and emphasise the value of proofs within simple puzzles and recreational mathematics where proofs are often much easier to come by — as illustrated by the efforts of Jennifer Dunham and Mary Watson.

Another reaction is to emphasise the element of play and playing around. Even professional mathematicians when faced with problems at their level of challenge and difficulty do not immediately spot the imaginative steps that they need for a proof: quite the opposite, *everyone* finds the exercise of imagination difficult and demanding.

This emphasises the scientific aspect, where playing includes experimenting, and looking out for equal segments, equal angles, collinear points, concurrent lines and so on.

In the next figure, a parabola is defined in the traditional manner as the path of a point which is equidistant from a fixed point, the focus, and a fixed line, the directrix. The tangent has been drawn at P and then the focus joined to Y, the foot of the perpendicular to the directrix. The tangent seems, by simple observation, to be the perpendicular bisector of FP: it is. What's more, if the tangent cuts the axis of the parabola at Q, then FPYQ looks like a rhombus, which is also true.

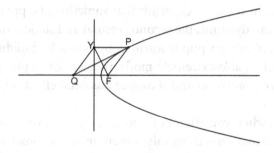

Simple properties of a parabola

However, these simple properties are also open to relatively simple proofs. By a dynamic argument (there is also a static argument) the point P is moving away from the focus and the directrix line at the same rate, and so ∠FPQ = ∠YPQ. Since FP = PY, ΔFPY is isosceles and PQ is indeed the perpendicular bisector of FY. Also since PY is parallel to FQ, ΔYQF is the mirror image of ΔYPF.

These arguments much resemble moves in a game. The number of possible *and plausible* moves[§] is limited, and an effective choice of moves leads to the desired conclusion. To spot an effective chain of moves, however, requires some imagination, insight and the ability to look ahead.

Problema versus theorema

> A problem is a proposition in which something is required to be done. Then successful doing of this thing is termed the solution to the problem. Whenever and wherever investigation is in progress, problems present themselves for solution.
>
> (Loomis 1902:13)

This epigraph was written by Elisha S. Loomis who was famous in later years for *The Pythagorean Proposition* which collected together 367 proofs of the theorem. He was referring to the traditional distinction between *problems* and *theorems*.

The distinction goes back to the ancient Greeks and was still alive in the sixteenth century and later. 'A Probleme, is a proposition which requireth some action or doing' as it was put in 1570 (Billingsley 1570).

A theorem, in contrast, demands that something be proved. The first term is active and dynamic, the second cerebral and static. *Problem* fits the progressive emphasis on pupils learning by *doing*, by building and constructing, which is also extremely motivating, so it is a pity that this old contrast between *problema* and *theorema* has been effectively lost in current textbooks.

Of course, when you attempt to prove a theorem you also have to do something, and that something might even involve a construction — and calculation is also a kind of doing, so the distinction cannot be drawn too sharply, but neither should it be lost sight of. (It has also come into its own again with the arrival of geometry drawing packages for the computer.)

[§] One possible move would be to draw a chord at random to cross the parabola, and this can indeed be done in an infinite number of ways: but such moves are not plausible in the context of drawing conclusions about this particular figure. The perception that such moves are not plausible is a function of experience.

Page 48 quoted Charles Hutton from his *Recreations in Mathematics*. He referred there to problems, and sure enough almost all his examples are just that, *problems* requiring something to be done, though he does warn readers that he has included a few *theorems* that require to be proved.

Elementary geometry, practical and theoretical by Godfrey and Siddons included several *problema*, or constructional puzzles. These two are from their first section on experimental geometry:

> To draw a parallel to a given line QR through a given point P by means
> of a set square and a straight edge.

> To draw through a given point P a straight line perpendicular to a given
> straight line.

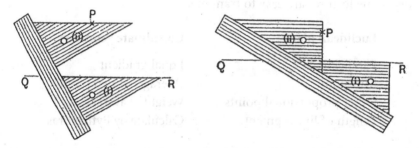

Solutions with a set square and a straight edge

These puzzles especially appeal to me because they are practical, they give pupils the chance to experiment and play constructively, and the first points to the properties of corresponding angles (Godfrey and Siddons 1903:36–37).

This on the other hand, is more of a pure puzzle but no less motivating for that, and also much harder if you take the qualification 'smallest' seriously:

> What is the smallest number of lines that you need to draw with a ruler
> or compasses in order to construct a square?

(The question is ambiguous: it could mean with a ruler only, with compasses only, or using both ruler and compasses, as necessary.)

The claim that your construction is a square might be doubted, as might the claim that the construction cannot be performed in fewer moves: both these claims are potential *theorema*, open to proof.

Euclidean geometry and coordinates

Starting with the idea of the Euclidean plane naturally divided up into a square or triangular grid of identical units is an open invitation to go a bit further and introduce the use of co-ordinates — but co-ordinates will then be motivated by being a simple extension of an earlier idea.

The Euclidean plane and the co-ordinate plane model each other: all the common features, concepts, object and properties of each can be translated into the other. They are therefore examples of *isomorphic* structures. Some features are easy to translate.

Euclidean plane	Coordinate geometry
Parallel lines	Equal gradient
Mid-point	'Average' of two points
Other proportional points	Weighted averages
Length of line segment	Calculate by Pythagoras

Others, such as angles, are harder, illustrating how some features are better handled in one model rather than the other — an idea that pupils should already appreciate through understanding that fractions are sometimes more suitable than decimals, but sometimes it's the other way round. The first calculation is much easier to do, with or without a calculator, than the second:

$$7/24 \times 6/13 = 7/52 \qquad 0.2916666... \times 0.461538461... = 0.1346153 ...$$

Proof using co-ordinates

It is now easy to prove many basic properties of simple shapes just by calculating co-ordinates, either with actual numbers, or using algebraic letters. Beginners can use actual numerical co-ordinates which exactly match their figure. Since this is only one specific case, it is necessary to explain — or prove! — that the argument for the specific case would indeed apply to any other choice of figure and co-ordinates.

More experienced 'players' (but this is a very big jump in difficulty) can use literals and a general figure.

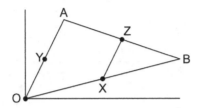

Proof by calculating mid-points

A basic tactic or method is to calculate the midpoint of the segment joining two points as the 'average' of the two points. Here $OZ = OX + XZ = 1/2OA + 1/2OB = 1/2A + 1/2B$ if A stands for its coordinate pair.

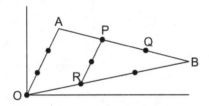

Thinking of 'vector' moves

The tactic here is to consider the 'vector' moves that you need to make horizontally and vertically on the underlying tessellation to move from one point to the other. The next step from this tactic is to use weighted averages: the point P, one-third of the way from A to B, can be thought of as being reached by the moves, $OP = OR + RP = 1/3\ OB + 2/3\ OA$. In weighted averages, the coordinates of P, which is twice as close to A as it is to B, are $(2A+B)/3$.

As usual, stronger pupils can exploit this to explore and prove much harder theorems, such as Menelaus and Ceva but, as always, this step upwards increases the difficulty enormously. They can also explore linear functions. In the same figure, the value of any linear function at P is the same weighted average of its values at A and B.

9

The Scientific Approach
to Mathematics

Mathematicians have always experimented, observed, and then drawn conclusions. Philip Davis asks,

> How were the theorems of triangle geometry discovered? I can only conjecture that as with much of mathematics, it emerged from long hours of 'playing around'.
>
> (Davis 1995:206)

The counting numbers also invite 'playing around' and conjecture, so it is no surprise that according to G.H. Hardy,

> The theory of numbers, more than any other branch of pure mathematics, has begin by being an empirical science. Its most famous theorems have all been conjectured, sometimes a hundred years or more before they have been proved; and they have been suggested by the evidence of a mass of computation.
>
> (Hardy 1920:651)

Needless to say, 'playing around' is a very enjoyable and motivating activity, and spotting patterns in a mass of data is especially rewarding: it certainly provides the 'kick' that Hardy used to explain the attractions of puzzles.

The traditional picture of mathematics saw it as a *deductive* science modelled on Euclid in contrast to the *inductive* natural sciences which relied on experiment and hypothesis and lacked the certainty of maths. This was the typically Victorian view of the scientist T.H. Huxley, but in an address in 1869 to the British Association for the Advancement of Science, Joseph Sylvester disagreed:

> [Huxley] says 'mathematical training is almost purely deductive. The mathematician starts with a few simple propositions, the proof of which is so obvious that they are called self-evident, and the rest of his work consists of subtle deductions from them' ... we are told [by Huxley] that 'Mathematics is that study which knows nothing of observation, nothing of experiment, nothing of induction, nothing of causation. I think no statement could have been made more opposite to the undoubted facts of the case' ...

Sylvester went on to claim that mathematics was extremely creative, repeatedly introducing new ideas and methods,

> from continually renewed introspection of that inner world of thought of which the phenomena are as varied and require as close attention to discern as those of the outer physical world ... that it is unceasingly calling forth the faculties of observation and comparison, that one of its principal weapons is induction, that it has frequent recourse to experimental trial and verification, and that it affords a boundless scope for the exercise of the highest efforts of imagination and invention.

Sylvester added in an appendix that 'Induction and analogy are the special characteristics of modern mathematics' (Sylvester 1904:v.II). We might add that game-like mathematics also gives scope for 'great imagination and invention', as we have seen in discussing the scale from 'pure' creativity to methods and then techniques, but we can agree also that mathematics has an extremely strong scientific aspect which pupils can enjoy exploring.

Mathematics is not 'A' science any more than it is 'A' game but it does *resemble* science: mathematicians do experiments (for example, a calculation, a drawing) collect data, form hypotheses, and test them — but being mathematicians, they then hope to *prove* that their hypothesis is true which scientists can never do — a general concept about maths and science that pupils should learn to appreciate.

The great Gauss is a happy example. He was a calculating prodigy who retained his ability into adulthood, and once replied to a question that he made all his discoveries through 'through systematic, palpable experimentation'. As a Mathematical Association report put it in 1919,

> Mathematics gives scope for inductive as well as deductive thought. Geometry in some of its aspects has many of the characteristics of a physical science, and lends itself largely to inductive processes; opportunities for induction, intuition and imagination are among the essential merits of mathematical study.

(Mathematical Association 1919:7)

They then recall one of the most famous of all examples of scientific induction in mathematics: Newton's discovery of the general binomial theorem for fractional indices (*Ibid.*:8).

Erroneous inductions

> Some branches of mathematics have the pleasant characteristic that what seems plausible at first sight is generally true. In (analytic prime number) theory anyone can make plausible conjectures, and they are almost always false.

(Hardy 1915:18)

Looking for hypotheses can be a very motivating experience, and the fact that pupils will sometimes form hypotheses that are false, is no objection: this is a commonplace in mathematics as it is in science. How convincing are the following patterns?

$$333667 \times 1113 = 371371371$$
$$33336667 \times 11113 = 371137113711$$
$$3333366667 \times 111113 = 371113711137111$$

Compare, $333667 \times 2223 = 741741741$

Most mathematicians would feel that they were totally convincing but only because of their past experience of patterns *of this type*. Someone unfamiliar with such patters might feel less confident.

What would a proof add in the first case? Certainty? Maybe not, but illumination, yes, which might lead to a wealth of other properties of the same type, based on whatever underlying principle the discovery of the proof reveals.

How suggestive is this single fact?

$$17^5 = 1419857$$

The number 1419257 looks surprisingly like the decimal period of 1/7: 142857. Is there a connection here or is this a 'pure coincidence' with no deeper significance?

Inductions based on drawing geometrical figures also tend to be very solid, despite Daniel Schultz's counterexample (pages 148–9).

The computer and experimental mathematics

It is plausibly no accident that two of the greatest experimenters in the history of mathematics, Euler and Gauss, were both brilliant mental calculators, walking computers, as it were. Today, fortunately, it is not necessary to be a calculating prodigy to benefit from giant calculations. Computers can be used to generate data and indeed to do much more.

Borwein and Bailey (2004:2–3) have used the term 'experimental mathematics' to mean methods that includes the use of computation for:

- Gaining insight and intuition.
- Discovering new patterns and relationships.
- Using graphical displays to suggest underlying mathematical principles.
- Testing and especially falsifying conjectures.
- Exploring a possible result to see if it is worth formal proof.
- Suggesting approaches for a formal proof.
- Replacing lengthy hand derivations with computer-based derivations.
- Confirming analytically derived results.

The difficulty of proof

The very fact that so many theorems have been conjectured long before they were proved suggests that even top professional mathematicians find scientific induction much easier than proof, so of course pupils will suffer from the same limitation. Therefore we should not expect pupils to prove everything they can discover, and they should be taught to appreciate the difference between the ease of pattern-spotting and the difficulty of proving: that does not mean that proof is not central, only that it is relatively difficult.

Experiment and aesthetics

Maths-as-science does not have quite the kick or buzz of game-like maths-with-proof but it does have powerful rewards: the results of experiment can be not merely surprising, but amazing and astonishing, as well as mysterious and challenging. They can also be paradoxical or seemingly impossible: indeed, this is typical of many breakthrough experiments in the sciences. They initially provoke incredulity, but if and when the experimental result is confirmed, they lead instead to rethinking current theory and so on to new theories.

Experiments can also, of course, reveal patterns and structure, or suggest invariance when the input to the experiment varies but output does not, as well as suggesting analogies and generalisations.

Thinking of a hypothesis to explain a set of data or an observation can give a buzz, even if the hypothesis turns out eventually to be false — and if it passes your tests and seems to be true then you can switch away from to experiment to proof and try to confirm your conclusion.

Investigations

> Nothing is as motivating as pleasure and this opportunity (investigations) for pupils to experience pleasure within mathematics was considered crucially important if more people were to pursue mathematics to a higher level. This argument does not appear to have succeeded and investigations, per se, are now recognised to be insufficient bait.
>
> (Rodd & Monaghan 2002:74)

The investigations' movement was the other mathematics education movement of the second half of the twentieth century, following the Modern Mathematics Movement. It was nothing like as damaging as the Modern Mathematics Movement, but neither was it a long-term success, and it has long since been steadily fading.

Why? Why did a movement that insisted that pupils should be doing mathematics — a most admirable goal! — first falter and then fail? One clue is in this striking fact: professional mathematicians, both pure and applied, seldom if ever talk of 'investigations' as a noun. They talk of solving problems. Of course, when attempting to solve a problem they will explore — investigate! — the problem, as they seek for the clues, the data, the ideas, the analogies, the partial and temporary inferences that they hope will lead them to a solution — but they describe all of these activities as (attempting) to solve the problem!

We have quoted Halmos on problems as the heart of mathematics. We could also quote Jean Dieudonné *eminence gris* of Bourbaki: 'The history of mathematics shows that a theory almost always originates in the efforts to solve a specific problem.' Would that teachers and maths educators would be satisfied with such a simple — but also profound and challenging — goal!

When certain keen and enthusiastic maths teachers and educators proposed that pupils should DO mathematics by DOING investigations they were marching against the behaviour of professionals. Why? In its origin the movement was less limited. Here is an account from 1969:

> It is the aim of this project to develop sixth-form (sic) work in mathematics in which a major part of the activity is the investigation by individual pupils of substantial open problems ... formulating problems, solving them, extending and generalising them.
>
> (ATM 1969)

The article concluded that, 'the kind of work exemplified above might be described as research-type activity at the pupils' level'. The same sixth form project gave a very similar account a year later. However,

> This account of investigations is a fair description of professionals doing mathematics. However: the activity described under 'Investigations' is what research mathematicians call *problem solving*, while the description

of problem solving describes only the routine solution of a problem using established techniques ... It does *not* describe what research mathematicians mean when they talk about problems. (1985:39)

Cockcroft contributed to the confusion by making an unfortunate link between problem solving and applications:

> Mathematics at all levels should include ... problem solving,
> including the application of mathematics to everyday situations; investigational work. (Cokcroft 1982:243)

> The ability to solve problems is at the heart of mathematics. Mathematics is only 'useful' to the extent to which it can be applied to a particular situation and it is the ability to apply mathematics to a variety of situations to which we give the name 'problem solving'. (*Ibid*.:249)

No, it isn't, or rather, it shouldn't be. Cockcroft wrote as if mathematics is always applied and as if pure mathematicians do not solve problems.

The degeneration of the idea of investigation

The claim that 'problem' should not be interpreted in a narrow sense — and especially not as an exercise problem from a textbook! — is fair enough, but that is no reason for swinging from one extreme to another. Eric Love wrote that,

> In the ATM writings of (the 1960s) it is clear that the people concerned were not only advocating students working at their own mathematics, but they were reacting against the idea that problems must be well-defined in a narrow sense ... There is talk of 'problem situations', of 'starting points'. The aim was always to help students make mathematics their own ... There is no mention of 'doing an investigation' ...
>
> (Love 1987)

But in professional mathematics there is no mention of *starting points* or *problem situations*: there are just problems, of many different types and styles. We might detect here a determination NOT to be definite, NOT to be precise. This was indeed a tendency among many promoters of investigations. One maths educator took this attitude to this bizarre extreme: 'Interesting problems are not clearly defined: clearly defined problems are not problems.'

The trouble with this claim is that, once again, it contradicts the use of the term 'problem' by professional mathematicians, and indeed most teachers. It is also contrary to common sense: Fermat's Last Theorem as posed by Fermat was perfectly clear. Was it therefore un-interesting? The suggestion is absurd. What about Hilbert's 23 problems famously presented at the 1900 International Congress of Mathematicians? They can be checked on the internet where they are readily available. They were all formulated clearly and precisely, which is not to say that in trying to solve them, mathematicians did not change old ideas, or come up with new ideas or reinterpret the original questions. Were Hilbert's problems lacking in interest?

Professional mathematicians not only find clearly defined problems interesting, but they tend to clearly define problems that interest them. Of course, it does not follow that they approach a problem in a narrow, unimaginative way, obsessively fixated on only one closed outcome. Far from it!

These strikingly unprofessional uses of the term *investigation* had other consequences. According to, *GCSE: A Guide for Teachers of Mathematics* (1986a:26)

> The term 'mathematical investigation' has come to mean a type of work whose value lies more in the activity of solving the problem than in the solution itself.

If so, then so much the worse for investigations. Far from the process of doing mathematics being sufficient reward in itself, the role of final goals as well as the medium-term goals that the process of problem solving creates are inextricably intertwined.

The idea that pupils doing art, playing tennis or playing a musical instrument, climbing a mountain or writing a story, would find the process more important than the final outcome, is extraordinary. The two aspects are more often than not inseparable, in the execution and the judgement, cognitively and affectively.

In particular, the rewards of a problem are often maximised when the pupil reaches a point — compare the climber standing on top of the peak or the artist standing back to gaze on the completed work — from which the previous efforts and failures and successes, can be viewed from a perspective that was not previously available.

We have already met the 'billiards' puzzle and Shamshad Ersan's proof. The idea that Shamshad got no additional 'kick' out of proving his conclusion by a sound argument is once again absurd. It suggests he would have got the same satisfaction if he had NOT proved his conclusion, or indeed not come to any conclusion at all, whereas in the real world, although professionals often have to accept failure — and so must pupils — they also hope for the deep satisfaction that comes from success attained after so much effort.

We may add — recapping — that proofs are associated with beauty and aesthetic reward: if 'the solution itself' is of no special value, then the role of aesthetics in mathematics is devalued.

Extensions

So determined were many proponents of investigations to emphasise the vagueness of the situation and its 'open' quality, that they created a concept that actively undermined the idea that any problem was ever solved. This was the idea of *extensions*.

It is right that we should undermine the naive idea that a problem once solved has nothing more to offer — professional mathematicians know better — but extensions have their own dangers: an emphasis on extensions can reduce motivation, by suggesting that whatever the pupil achieves is never enough.

If the 'kick' that pupils get from solving a problem is weakened, then they lose motivation. If a pupil has reached a solution to a problem, but the teacher is an extension-maniac then a voice whispers in the pupil's ear, 'What about …? What if …? Why don't you …?' and the pupil's sense of satisfaction at solving the problem is diminished.

Eric Love concluded his quotation above by suggesting that,

> In the last five to ten years 'investigations' have become institutionalised —
> as part of formal requirements for assessment of courses ... Such a
> development is a typical one in education — the often commented-upon
> way in which originally liberating ways of working become formalised
> and codified, losing their purpose ...
>
> (Love 1987)

Rodd and Monaghan offer a similar explanation:

> Furthermore, for investigations to be 'integral' to the curriculum they
> had to be assessed. So exam boards set clear grade descriptions resulting
> in 'investigating' becoming ossified and unimaginative. The fact is that a
> real open-ended mathematical quest does not fit nicely into lesson-
> allotted time spans of 35 or 50 minutes, let alone an examination.
>
> (Rodd & Monaghan:74)

The DPG model of investigations: maths-as-science

The distortion of the nature of mathematical activity through an over-
emphasis on 'investigations' had another consequence: mathematics was
presented as unduly science-like. Here are some illustrative comments:

> Investigations provide the experience of collecting and organising data ...
> (SPLASH May 1979)

> A good answer (on the SMILE Mode 3 O-level) ... will show ... evidence
> of the observation of any patterns in the results, the formulation of rules
> which describe any generalisations discovered ... (SPLASH June 1980)

> Investigations necessarily entail that children collectively acquire, sort
> and classify data, then hypothesise, test and predict on the basis of that
> data. (Mertens 1987:65)

In *Problem Solving and Investigations* (1986/1993) I argued that the investigations movement as it had developed promoted a data-pattern-generalisation (DPG) model of research-type activity. This DPG model emphasises

pattern spotting and downplays proof and every other aspect of mathematics: therefore, although the original idea of investigations was that pupils would be behaving more like young mathematicians, the result of the ossified DPG paradigm is that they behave LESS like mathematicians, rather, they are reduced to being young scientists — which is only aspect of doing mathematics. As Dylan Wiliam put it:

> What I find particularly saddening about the DPG paradigm is that it reduces mathematics to an empirical discipline. Once the 'rule' has been discovered, there is no need to go any further. To my mind, this is not mathematics, any more than Hilbert's axiomatic rule game is mathematics. Mathematics is as subtle, as complex, and as fascinating as it is because the empirical aspects and the logico-deductive aspects are like the warp and the weft of a fabric: take away either and you have almost nothing left.
>
> (Wiliam 1993)

Polya and the origins of the DPG model

Historical research uncovers a possible source of this bias within maths education (1986a:12–13). George Polya's 1945 book, *How to Solve It* is a cornucopia in which induction features as only one among many other heuristics. *Mathematics and plausible reasoning: Vol. 1, Induction and analogy in mathematics*, and *Vol. 2, Patterns of plausible inference* (1954) is not so even handed, nor was it intended to be. The first volume has seven chapters on induction and 'Further kinds of plausible reasoning' are only the subject of the final chapter.

Ten years later, the two volumes of *Mathematical discovery: on understanding, learning and teaching problem solving* (1965) were aimed at teachers. As a whole it could not be interpreted as promoting a DPG model and yet the very last chapter is 'Guessing and the scientific method'. Starting, 'Non-mathematical induction plays an essential role in mathematical research', which is quite true, Polya calls for 'Research problems on the classroom level' and presents two imaginary dialogues between a teacher and a class to 'show how a good teacher can offer something approaching the experience of independent inquiry even to an average class by choosing appropriate problems … ' (*Ibid.*:144).

These dialogues are bizarre because in both, data are generated and inductive generalisation made from the data. They are quite untypical, therefore, of most of Polya's other examples. Only at the end do any of his other heuristics appear. Polya continues, 'These remarks ... even as they stand ... can give students on the high school level a basic insight into the nature of *science*' (*Ibid.*:156; italics added for emphasis). Here is an obvious invitation to construct DPG investigations.

Ideological roots of the investigations movement

There is yet another connection between Polya and the investigations movement. Polya's second example of research problems on the classroom level (*Ibid.*:143) is Euler's formula for connecting the faces, edges and vertices of a polyhedron and it was Polya who suggested to Lakatos that he use historical attempts to prove this relationship as the subject for his thesis, which became the book *Proofs and Refutations*, which had a tremendous influence in promoting a science-like picture of mathematics in which every mathematical proposition is dubitable.

The title, ironically, is a play on the title of Popper's *Conjectures and Refutations*. If Lakatos had followed his teacher, Karl Popper, and subjected his thesis to the severest, rather than the mildest possible, test, it would have seemed nothing like so powerful and its effect on mathematical education would have been very different.

Lakatos needed an example which could be presented as based on the fallibility of mathematical proof — although it would inevitably be actually based on failure of conceptualisation.

Lakatos claimed that *informal* — note that essential qualification — mathematics develops like natural science. Conjectures are made and proofs sought, counterexamples are discovered and the proof/theory is changed, and so on. He argued this thesis in the body of the book by focusing on one development, the history of Euler's relationship for polyhedra, presented in dialogue form. In Appendix 1 he adds, 'Another case-study ... Cauchy's Defence of the 'Principle of Continuity''. He initially, and explicitly, developed this claim as an argument against formalist philosophies of mathematics, and as a pair of dazzling counter-examples to formalist theses, he succeeds brilliantly. Unfortunately, he was not satisfied. The subtitle of

Proofs and Refutations is *The Logic of Mathematical Discovery*. Note the definite article, and the lack of any other qualification. He is not talking about one aspect of mathematical discovery, but of mathematical discovery in general.

He commences his second example, discussing Cauchy, thus: 'The method of proofs and refutations is a very general heuristic pattern of mathematical discovery.' In this sentence, note, he does not qualify 'mathematical discovery' and indeed he writes just below, 'There is a simple pattern of mathematical discovery — or the growth of informal mathematical theories.'

'Or'? Are these alternatives? Since when has discovery in mathematics been limited to the growth of 'informal mathematical theories'? Vast tracts of 'mathematical discovery' certainly do NOT consist only of the development of such 'informal theories'. Lakatos produces two superb counterexamples to formalist generalisation. On the other hand, he apparently wants to generalise from these examples to 'mathematical discovery' in general — an absurd aim and a dud induction if ever there was one. His mentor, Popper, must be turning in his grave.

The problems Lakatos chooses to analyse all involve questions of definition, that is, processes involved in the very construction of game-like mathematics, but Lakatos, who has no conception of game-like mathematics does not realise that this is an important distinction.

The search for effective definition can be interpreted as especially scientific. When Cauchy had sorted out the convergence of convergent series to his satisfaction and nineteenth century mathematicians turned their attention to divergent series their investigations can be seen as highly scientific. There are many ways to define the summability of a divergent series. Which will be the most effective? By what criteria? Which definitions are equivalent? What desirable properties do the various definitions possess? The result was not proof that such-and-such a definition was correct — definitions cannot be the result of proofs. Rather, a scientific judgement was made that certain definitions were effective and useful. This is how mathematicians eventually came to settle the question of proving Euler's relationship — but it is NOT a model for most mathematical research which takes place within already game-like structures (1988a).

'Anti-authoritarianism' tendencies

> The influence of these developments in mathematics has been strongly
> reinforced by the claims of some mathematics educators, inspired in part
> by the work of Lakatos, that deductive proof is not central to mathemat-
> ical discovery, that mathematics is fallible in any case, and that proof is
> an authoritarian affront to modern social values and even hinders learn-
> ing among certain cultural groups.
>
> (Hanna 1996:1)

There are philosophical and psychological 'reasons' for some teachers to
downplay the role of technique. To strongly libertarian teachers, the very
idea of proving a proposition once-and-for-all may be disturbing while
the discipline involved in developing technique may also be off-putting,
even anathema.

An amusing, but significant, example occurred at the meeting of the
Philosophy of Mathematics Education group at the British Congress of
Mathematical Education in 1991. A remark by Paul Ernest, the leader of
the group, prompted me to ask him if he really thought the statement
13 + 17 = 30 is 'dubitable' and he replied, 'Yes'. I noticed later that Eric
Smith also records that,[‡] 'Paul Ernest stated that the mathematical statement 2 + 2 = 4 is fallible' (Smith 1996:81).

Unfortunately, libertarian biases and prejudices are a hindrance to the
philosophy and psychology of mathematics education, not an aid. It is true
that Euclid's many theorems were based on less than perfect foundations —
but when Hilbert reworked Euclid's foundations not a single theorem
fell over. *Pace* Paul Ernest there are vast numbers of mathematical theorems that only an ideologue with a position to support would describe as
'dubitable'.

Likewise, the many resemblances between mathematics and abstract
games are as old as the Greeks and beyond, and are not at all affected by
modern ideologies.

[‡]This was also a 'personal communication' but it is printed in a book edited by Ernest with
no editorial qualification so presumably does not misrepresent his views.

Game-like AND scientific

> I would like to see a move away from the culture which regards geometry
> (and all of mathematics) as an experimental science, in which general
> truths emerge as mysterious laws of nature. The point of proof at school
> level is that it provides *explanations*.
>
> (A professor of applied mathematics: Royal Society 2001)

Euler and Ramanujan were both famous for their brilliant inductions
and for their brilliantly creative game-like manipulations. Euler's discovery and proof of his pentagonal number theorem is a wonderful example
of both (2007:95–98). (George Polya identified it only as a brilliant
example of induction, totally overlooking Euler's brilliance at game-like
transformations.)

It is no accident — but rather a consequence of their formalisation —
that not only elementary arithmetic and algebra but also a great deal of
elementary geometry and number theory, tend to be especially game-like
and to be especially open to scientific investigation, which also explains
why they are recreational for so many amateurs.

Mathematics as science-like must be adequately stressed. However, there
is a danger in swinging from a one-sided emphasis on (naive ideas of)
deduction to an equally extreme emphasis on the scientific aspect — in each
case, I will add, paying too little attention to a third aspect, the perceptual
and interpretive.

The game-like aspects of mathematics are at least as motivating as the
scientific — as well as distinguishing mathematics from science. We cannot
maximise motivation if the scientific aspect is over-emphasised and the
game-like, including proof, are played down: least of all if this is done for
purely ideological reasons which do not reflect the actual experiences of
mathematicians.

10

Mathematics as Perception

Very complex processes of perception lie behind game playing and science and indeed all our activities. The game player is said to have more-or-less quick 'sight of the board' as well as seeing — or failing to see — structural connections between the features of a position which guide him in selecting his moves.

The scientist not only makes observations during experiments, but then 'sees' patterns, analogies, connections, structural features, which can be built into a scientific model. Richard Feynman emphasised the importance of 'seeing apparently different phenomena as essentially the same', and the mathematician needs the same ability (Feynman 1964:v.1:28). So do pupils, but unfortunately they vary greatly in their ability to 'see'.

We know that mathematics is about pattern and structure, which are often thought of as very visual phenomena. However, mathematics also involves perception in much more subtle ways. When Shamshad Ersan created his solution to the billiard table problem, he was 'seeing' the movement of the ball as a combination of horizontal and vertical motions which are partly independent. When 'Gary' solved the problem of the two overlapping maps he used his imagination to create a grid that was 'not there' in the original figure.

What can pupils see?

The illustration below is a familiar puzzle which might seem to have no connection whatsoever with mathematics, but that is not so.

How many hidden animals are there? (Wells 1988f:20)

Where are the animals in this scene? Why are they so difficult to spot?
Would you spot them at all, if you were not told that they were there?

We often expect pupils to spot parallel lines, right angles, or similar triangles in a complex figure, but some will find this much easier than others.

Some pupils will readily see, without drawing any actual lines, that in the next figure you can join A and B with multiples of the vectors (4,1) and (2,3). This is perhaps more likely if they play chess, because this kind of movement is close to the knight's move.

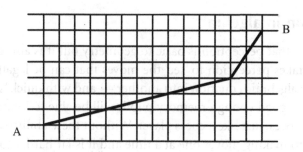

Combining vectors to reach a goal

They may even see that this can be done in several ways. Other pupils will need to draw the actual lines, and then make mistakes in doing so. Some pupils, likewise, may spot at once that the three crossing points in the middle of this figure lie on a straight line, but others may not:

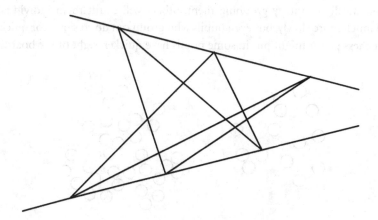

How do you 'see' the crossing points?

Individual differences in perception are important in geometry, algebra and every part of mathematics. Pupils in their ability to visualise and to 'look ahead' visually vary greatly (and with the kind of visualisation involved) all the way from very poor to people like Robert Reid, a well-known puzzlist, who can create novel dissections of dodecahedra 'in his head', only constructing the actual physical model later.

Perception in algebra

It is obvious that perception is basic to geometry but it is also essential to algebra. It takes perception to 'see' the moves that can be legally made to change one algebraic equation into another — and with quick 'sight of the board' plus a good enough memory pupils will be able to look ahead and plan their moves, just like a chess player. Without these abilities, the player is reduced to making moves one at a time and it is far harder to see which moves lead to good outcomes and which are dud.

How many circles? How many triangles?

This test is extremely crude but even so will show up strong differences between pupils: how many circles are there in these figures? Pupils vary greatly in how quickly they can count these circles and triangles accurately. Some will automatically count by grouping them, others will count them individually and much more slowly, but even pupils who group can do so slowly or quickly. As a chess player might put it: some pupils have quicker 'sight of the board'.

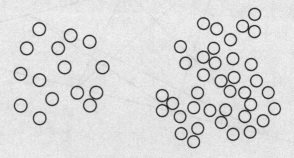

How do you count the circles?

(Continued)

(Continued)

How many triangles are there in this figure? This is harder, because some triangles are composed of other triangles. It also shows striking differences between pupils, suggesting differences in pupils' ability to scan geometrical diagrams and pick out similar triangles, equal angles, and pattern and structure in general.

 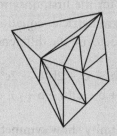

How many triangles in these figures?

It takes a small degree of insightful perception and 'look ahead' just to write down this product without any intermediate steps:

$$(x + y + 1)(x - y - 1) = x^2 - y^2 - 2y - 1$$

Symmetry also features in algebra, though it is often hidden. It is not superficially clear that this quadratic is symmetrical. The symmetry is seen more clearly when it is factorised:

$$x^2 - 5x + 6 = (x - 2)(x - 3) = (x - 2\frac{1}{2} + \frac{1}{2})(x - 2\frac{1}{2} - \frac{1}{2})$$
$$= (x - 2\frac{1}{2})^2 - (\frac{1}{2})^2$$

The same is true of a cubic, as we saw earlier. If we take a symmetrical expression and treat it symmetrically, then the result will be symmetrical also:

$$x + y$$
$$(x + y)^2 = x^2 + 2xy + y^2$$
$$(x + y)^3 = x^3 + 3x^2y + 3y^2x + y^3$$

And so on. Sometimes however we find symmetry where we might not have expected it.

If $\qquad$ $a + b = c$
then $\qquad$ $a^2 - b^2 - c^2 = 2bc$
and $\qquad$ $a^4 + b^4 + c^4 = 2b^2c^2 + 2c^2a^2 + 2a^2b^2$

How is it possible for the first, unsymmetrical, equation to turn into the final extremely symmetrical equation (1980–1983:8:13)? The answer appears when we factorise the difference completely:

$$a^4 + b^4 + c^4 - 2b^2c^2 - 2c^2a^2 - 2a^2b^2$$
$$= (a + b - c)(a - b + c)(-a + b + c)(a + b + c)$$

The cube roots of unity show symmetry in a different way. Calculated by algebra, they seem somewhat idiosyncratic and unsymmetrical, but show them on an Argand diagram and the image is symmetrical. The visual image shows features — roots are of equal modulus and equally spaced round the circle — that are not obvious in their original algebraic form.

Algebraic games and interpretation

By the standard *rules-of-the-game* as taught to millions of pupils, if,

$\qquad$ $4x = 2x + 10$
then $\qquad$ $4x - 2x = 2x - 2x + 10$
and $\qquad$ $2x = 10 \quad$ and $\quad x = 5$

However, the equation $\qquad$ $4x = 2x + 10$
can also be thought of as $\qquad$ $2x + 2x = 2x + 10$
so, 'by looking', $\qquad$ $2x = 10 \quad$ and $\quad x = 5$

The first solution is game-like, the second involves *an interpretation, a way of seeing*. However — and this is typical — if pupils are taught to split the term as in the second example, then this could be seen as a purely game-like move. As we have already emphasised, all game-like moves involve perception.

Geometry providing metaphors for algebra

Geometry does more than appeal directly to the eye which can perceive in two and three dimensions and almost instantly detect (thanks to the organisation of the human visual system) such features as smoothness, boundaries and straight lines. Geometry, or rather geometries since there are many, provide superb motivating metaphors and analogies for other parts of mathematics. Thus T.J. Fletcher in his book, *Linear Algebra through its Applications* pauses at Ch. 2 to devote an entire chapter on three dimensional analytic geometry, explaining,

> ... it is the purpose of this chapter to assemble a relatively small number of powerful geometrical images for reference later in other contexts.

Why? Because such geometrical images, including ideas of the axes of symmetry of an ellipse or ellipsoid, are helpful to the further development of linear algebra and matrices. Compare the author of this quotation who is talking about differential equations and their *gradient pictures*.

> The geometric study of differential equations attempts ... to visualize the behavior of the entire system at once. The first picture which allows this heroic perspective is the gradient picture. Once glimpsed this description reappears in applications to all the sciences and whenever it applies it displays its power of organizing the system, with a single vision. When compelled by the complexities of more sophisticated systems to move beyond it, you leave with regret and never abandon it completely. *Always you look first for those aspects of the system where the intuitions of that initial vision still endure.*

(Akin 1993:1; italics added for emphasis)

We can also compare this quote from the statistician M.G. Kendall in *A course in the geometry of n dimensions* (1961):

> It is not easy to develop a comprehensive theory of statistics without introducing n-dimensional geometry at a fairly early stage.

Applied mathematicians have always valued the geometrical aspects of mathematics because they know that its ideas and imagery are so often an invaluable aid to understanding and problem solving, as this quote illustrates:

> For some time I have felt there is a good case for raising the profile of undergraduate (sic) geometry. The case can be argued on *academic* grounds alone. Geometry represents a way of thinking within mathematics, quite distinct from algebra and analysis, and so offers a fresh perspective on the subject. It can also be argued on purely *practical* grounds. My experience is that there is a measure of concern in various practical disciplines where geometry plays a substantial role (engineering science for instance) that their students no longer receive a basic geometrical training. And thirdly, it can be argued on *psychological* grounds. Few would deny that substantial areas of mathematics fail to excite student interest; yet there are many students attracted to geometry by its sheer visual content.
>
> (Gibson 2003:xi; italics in the original)

Geometry also has a powerful metaphorical role on a grander scale. It is much easier to surprise pupils with geometrical properties — such as three lines mysteriously concurring — than it is to surprise them through algebra, because they bring to geometry at least some intuitions from everyday life of what to expect in geometrical situations, and a little enjoyable experiment will give them a lot more of the same.

So geometry is a powerful ground on which to pose good problems, but equally, because elementary geometry can be so game-like, it is an excellent medium for learning to develop sound arguments.

In other words, geometry as a collection of striking objects with often equally striking properties, many of which are connected to each other, provides an especially vivid metaphor for other parts of mathematics, in which the objects and their relationships are, for young pupils, neither clear or intuitively easy to grasp.

Styles, perceptual and otherwise

We quoted G.H. Hardy earlier on the 'kick' that puzzles provide. Hardy then drew the conclusion that,

> The fact is that there are few more 'popular' subjects than mathematics. Most people have some appreciation of mathematics, just as most people can enjoy a pleasant tune ... There are masses of chess-players in every civilised country ... and every chess-player can recognise and appreciate a 'beautiful' game or problem. Yet a chess problem is *simply* an exercise in pure mathematics ...
>
> (Hardy 1941:86–88)

Hardy did not discuss the fact that millions of people do not play chess and many people actively dislike all abstract games. Moreover, though the rules of chess or draughts are very widely shared, even among chess players there are large differences in perception, and each player's understanding, insight and intuition are also strongly personal: these differences are linked to differences in style of play, in preferences for certain types of positions, and so on. There are also many idiosyncratic differences in the appreciation of mathematics:

> At the 1981 SMILE conference a speaker recommended for classroom use a very effective but also very crude method for quickly printing many units of a tessellation. He was enthusiastic but I was revolted. I love the precision and clarity of tessellations, while the care needed to construct a good tessellation, the insight into very simple geometrical principles required and the many small but very significant problems that pupils must solve on the way make a pedagogically very attractive and valuable task. Yet I do not doubt that the speaker finds great advantages in his approach. I am not arguing for one approach at the expense of the other — not at all. I am presenting the difference as typical of the differences that separate teachers. (1982:1)

Many cognitive styles have been identified — we shall not discuss them here — but their application to the classroom is controversial. However, even if it is hard to apply a cognitive style created in a different context to specific mathematics classrooms it does not follow that there are not cognitive differences that we should recognise and profitably take into account.

We have already considered the difference between a static and a dynamic approach to a topic. This is not a recognised psychological cognitive style but it is a stylistic distinction in the classroom. Some pupils are better than others at visualising dynamic transformations: those who have difficulty in visualising dynamic models will not be motivated and will not learn effectively by that approach.

We were walking through a field when R. noticed one particular flower and commented, 'It has ten petals in five pairs. *I worked it out once*'. The italics are mine. Because I am a very visual person, I could see more or less instantly that the petals came in five pairs. R. could only see this after a mental effort which took a significant time. Since R. will not always take such time, she will often not see what I see, and so structures which are there to me will be absent to her.

R. is relatively non-visual. However, when it comes to music, our situations are reversed. Her appreciation of music is much richer than mine, hence her comment that 'pop songs take a simple idea and treat it simply'. Not to me they don't, but my perception of music is itself very simple. I can well hear myself saying, while listening to a piece of music, 'That piece has five beats to the bar, I worked it out once ... '

This difference relates to the counting circles and counting triangles 'tests' described on pages 206–7 and might also be connected to Witkin's construct of psychological differentiation and field dependence/field independence. Unfortunately, although Witkin's research is 60 years old now, little has been done to relate his dimensions to mathematical perception and learning.

We might think that it is unfortunate that pupils differ so much in their ability to think visually. It is — but the other side of the coin is that they differ at least as much in their ability to think in abstract terms: take away the visual from elementary mathematics and what is left is even more difficult for most pupils. Yes, maths is a difficult subject!

Visual-versus-verbal is just one striking difference in perception and cognitive style. The Bourbaki group were very verbal. Felix Klein was a visual mathematician, who insisted that his students make models:

> (Klein) had a strong power of geometric visualization, and all his investigations were essentially governed by appropriate geometric pictures ... if you look at the drawings in his papers on automorphic functions, you will be astonished by the beauty of these figures, most of which are made up of very simple basic figures like triangles with curved sides. This beauty rests precisely on the fact that these figures illustrate the underlying mathematical relationships in an extremely simple and transparent way. Since Klein builds on these figures, all the results he derives possess that self-evidence which, as we said before, is the goal of mathematical research.
>
> (Krull 1987:50)

Anthony preferred that everything be explained in terms of rules. Soon after starting A-level he confided to me that he was pleased to find that he could understand the techniques being taught. Once, having asked me about integrating sin 2x and appearing to grasp the explanation perfectly well, he turned as he walked up the stairs, and gave me the rule for integrating sin ax having mentally turned the individual steps — the superior form of understanding from my point of view — into a rule which he prefers, an inferior form of understanding. Anthony came from a culture where rule-following is common, so perhaps he was only fulfilling a cultural stereotype but the effects were very real in the way he preferred to learn.

Adriaan de Groot in *Thought and Choice in Chess* (1965) made a distinction between players such as ex-world champion Max Euwe (a secondary mathematics teacher and amateur chess player) who were extremely 'logical' and those such as Tartakower, a grandmaster with a wild imagination and a capacity for fantasy at the chess board. This distinction is reminiscent of Liam Hudson's between pupils whom he described as either convergers or divergers (Hudson 1966, 1968). We might expect some of these differences to appear in maths pupils.

The next quotation illustrates a striking difference between two top professional mathematicians. The argument is between Michael Atiyah and Saunders Maclane, who recorded the incident:

> I adopted a standard position — you must specify the subject of interest, set up the needed axioms, and define the terms of reference. Atiyah much preferred the style of the theoretical physicists. For them, when a new idea comes up, one does not pause to define it, because to do so would be a damaging constraint. Instead they talk around the idea, develop its various connections, and finally come up with a much more supple and richer notion .... However I persisted in the position that as mathematicians we must know whereof we speak ... This instance may serve to illustrate the point that there is now no agreement as to how to do mathematics ...
>
> (MacLane 1983:53)

MacLane is being misleading. There never has been a uniform style of doing mathematics. From Archimedes to Newton to Hilbert and up to the present day, mathematicians have thought and worked in different styles. The work of Atiyah is more geometrical than that of MacLane who is an algebraist. This is Atiyah's view of geometry:

> Geometry is that part of mathematics in which visual thought is dominant whereas algebra is that part in which sequential thought is dominant. This dichotomy is perhaps better conveyed by the words 'insight' versus 'rigour' and both play an essential role in real mathematical problems.
>
> (Atiyah 2003:29)

We can reasonably assume that if Atiyah were (somehow) forced to do his research in the style and spirit of MacLane, and *vice versa*, each would been less motivated and their successes would have been fewer. Fortunately, professionals can choose, to a high degree, the approach to take. Pupils have less choice, and usually none.

Women and cognitive and perceptual styles

Although women and men overlap on their cognitive and emotional features, there is a good reason to think that their attitudes and abilities in mathematics show *stylistic* differences.

Women tend to score lower for psychological differentiation than men which is plausibly relevant to the perception of geometrical figures. Baron-Cohen has claimed that autism is a realisation of the 'extreme male brain' in which boys are keener on systems and girls on empathising (Baron-Cohen 2004). I don't entirely buy these claims but boys are more often than girls catagorised as nerds and seen as addicted to computers and mechanisms. This does not contradict Leone Burton's conclusion that girls are more at home with routines and situations *where they know what to do* (Burton 1990:.33c). In a rich and fascinating paper by Sherry Turkle and Seymour Papert, titled 'Epistemological pluralism: styles and voices within the computer culture' they wrote that,

> When we looked closely at programmers in action we saw formal and abstract approaches; but we also saw highly successful programmers in relationships with their material that are more reminiscent of a painter than a logician. They use concrete and personal approaches to knowledge that are far from the cultural stereotypes of formal mathematics.
>
> (Turkle & Papert 1992:48)

They relate these concrete approaches to Levi-Strauss's concept of the *bricoleur*, who 'constructs theories by arranging and rearranging, by negotiating and renegotiating with a set of well known materials', which they interpret as one style of thinking, among others, noting that even Nobel laureates relate to their materials in 'concrete and tactile' styles (Turkle & Papert 1992:49, 51).

They also suggest that many women, a larger proportion than of men, are likely to think in such concrete ways, writing that (1994a),

> Several intellectual perspectives suggest that women would feel more comfortable with a relational, interactive, and connected approach to objects (in contrast to) men with a more distanced stance, planning, commanding, and imposing principles on them.

We would like to adapt our teaching — as far as this is possible in very mixed classes — to the cognitive styles of our pupils. Fortunately, by emphasising different aspects of mathematics, from the perceptual to the scientific to the game-like, we can hope to motivate a wide variety of pupils and help them all to succeed.

.

Pattern and structure

Primary pupils have traditionally met structure very informally through visual patterns and making models. Abstract ideas of structure were first introduced by the Modern Mathematics Movement but they failed abysmally because their proponents took adult, professional concepts and attempted to impose them on children — contrary to every principle of sound teaching, of which the top-level professional mathematicians involved were totally ignorant — rather than allowing children to develop the central ideas through their own experience. This did incalculable damage to the cause of introducing pupils of all ages to ideas of pattern and structure.

> The art of mathematical proof often consists of finding a framework in which what one is trying to prove becomes nearly obvious. Mathematical creativity consists largely in finding such frameworks. Sometimes one finds them in the rich world of material objects, sometimes (and this is the highest form of creativity) one invents them. More often than not, one recognises that what one is interested in happens to fit into an already existing framework that was introduced originally for an entirely different purpose. (When a framework is used repeatedly in different contexts, it becomes a theory and is studied for its own sake.)
>
> (Kac & Ulam 1968:62; 1982:5)

Here the word 'framework' could be replaced by the words 'structure' or 'interpretation' or even by the phrase 'way of looking'. Such abstract structures are to the mathematician, via the problems in which they originate or to which they relate, strongly motivated, with a powerful affective content. They are vital tools with strong aesthetic features. Pupils are easily challenged by problems. How can their more modest efforts at solution be used to develop concepts of structure?

The structure within

The skeleton is a powerful analogy for introducing the idea that there are — often, sometimes — surprising and mysterious underlying patterns in an image or object. It is easy to spot the analogies between these two skeletons:

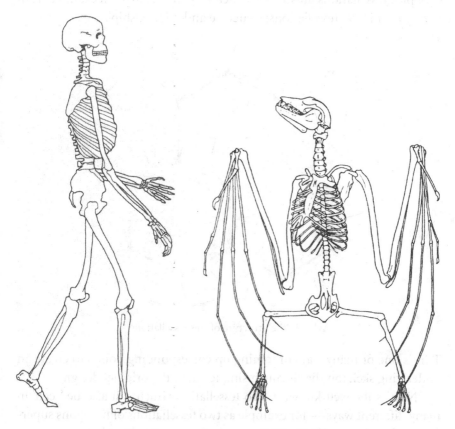

Spot the common structure (1988a:42–43)

Although a horse and a human being may look very different from the outside, inside they are both mammals (and so are elephants, giraffes and pigmy shrews) and their skeletons are 'obviously' sort-of-the-same.

Drawing and analysing tessellations

Complex tessellations illustrate the idea very well. Here is an example from (1987a) 'Unit 29: Tessellations: structure and relationship':

What is 'the' structure of this tessellation?

This is one of many ways of joining up corresponding points to create an underlying skeleton that is much simpler than the original design.

Next is the well-known Cairo tessellation which can also be 'seen' in many different ways — for example as two tessellations of hexagons superimposed at right angles, or as a pattern of squares each with the same pattern inside it:

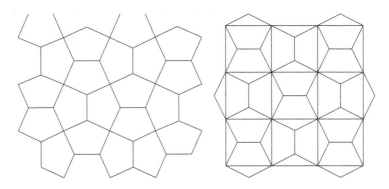

A structure for the Cairo tessellation

The idea of isomorphism and isomorphic problems

Mathematics is the art of giving the same name to different things.

What we must aim it is not so much to ascertain resemblances and differences, as to discover similarities hidden under apparent discrepancies.

(Poincaré 1930/2003:21)

Mathematicians are always on the lookout for structural resemblances because they know that so many exist, that they are incredibly powerful tools for problem solving and also extraordinarily beautiful, so the search for relations between structures can be intense and extremely exciting. As Gian-Carlo Rota put it, referring to some powerful modern theories:

There is an uneasy suspicion abroad that several subjects now considered to be distinct are really one and the same, when properly viewed.

(Rota 1988:141)

Children meet matching problems, matching situations — which they need to recognise are indeed matching — from the start of their primary schooling. The problems of adding five apples to three apples, and measuring the total length of two planks, 5 m and 3 m long, are 'identical'. The two problems match, when looked at suitably, meaning that you focus on the underlying arithmetic and not on the apples or metres.

Pupils meet such isomorphic problems many, many times — every time they do two addition sums, or three sums in adding fractions, or multiplying fractions or solving linear equations. Problems of the same 'type' have the same underlying structure.

So it might seem that such examples would be ideal to illustrate the idea of 'the same structure' but unfortunately 'spotting' such analogies is too simple — they are too abstract and too 'familiar' (unlike the analogy in Dali's painting) to be highlighted as examples of identical structure. As usual, extremely abstract simplicity is only suitable for much more sophisticated pupils.

The Metamorphosis of Narcissus by Salvador Dali

Three Essays on the Teaching of Mathematics incorporated a coloured postcard reproduction of Salvador Dali's painting, *The Metamorphosis of Narcissus*. Young pupils — meaning early secondary, average and below-average — find the full-size verion fascinating, not least because it is puzzling and *not* obvious: for most pupils, it only 'dawns on them' after a little while that the head and the egg 'match'.

'The reproduction ... of *The Metamorphosis of Narcissus* by Salvador Dali is a fine example which has the additional advantage of relating directly to another subject area. The hand holding the egg and the kneeling figure of Narcissus 'obviously' have the same structure, though pupils may take a little time before they 'see' this.

This one picture provides a splendid opportunity to use the language of structure. Each figure is made up of several parts and there is a simple relationship between them. The egg held by the hand corresponds to the head of Narcissus, the thumb is equivalent to the upper arm ... and taken as two wholes, the figure are isomorphic, they have the same structure.

This use of language is not mathematically very precise, but this does not matter. Quite the opposite, not only is it inevitable that such cases of 'isomorphism' will be tinged with everyday interpretations, but it is a good thing that this should be so. The words are being used, as it were, somewhat metaphorically, to bridge the gap between the pupils' everyday interpretations and more specifically mathematical interpretations ...

Although this particular painting will be unfamiliar to pupils, except by accident, this is not a case of unfamiliar content being used to teach new and strange ideas, because the painting is not content in that sense. Rather it is a striking illustration which pupils can easily appreciate and on the basis of which they can start to learn the significance of words like structure and isomorphism in a visually attractive and compelling way ...'

(1982a:14–15)

Isomorphic problems

The idea that many problems seem different but on closer examination are basically the same, is a deep one, and also an entertaining one. It is also powerful: spot the connection and solving one problem solves all problems of the same type.

In all areas of mathematics it is essential to be able to recognise the underlying type of a problem behind the surface camouflage. Much of the difficulty pupils have with word problems is that the underlying type is so well disguised by the verbiage that they become confused.

One example is the 'three pipes filling a bath with water' problem which my 96-year-old mother remembered with horror — but she was not embarrassed to admit that she once scored 2% in a school mathematics exam. This problem is a byword for uselessness, but that reputation is unfair, as Pendlebury (1897:264) makes clear:

> 299. If we know the time in which two or more persons can separately do a certain piece of work, we can easily find the time that they take to do the same when working together, supposing that each person still works at the same rate as when alone. Problems which involve the rate at which water or any liquid flows along a pipe are of a similar character. We will work out an example of each.

Just so: the problems are isomorphic, and since the first of the two at any rate is very practical, the method cannot be useless. Far from it, these puzzles also highlight an important general concept, that when dealing with speed, averages are often misleading. Here are some more examples of pairs of isomorphic problems, the first few from *Hidden Connections Double Meanings* (1988c:110–112):

1A   How many triangles are there in this figure?

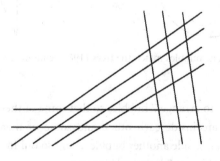

A nest of triangles

1B   In how many ways can a casting director choose a mother, a father and one child from two actresses, three actors and four children?

2A   How many numbers are equal to the sum of three of their factors which are all different?

2B   In how many ways can 1 be expressed as the sum of three different reciprocals?

3A   The sum of two numbers is 10. How big can their product be?

3B   What is the maximum area of a rectangle if the sum of its four sides is 20?

Bubbles and trees: a pair of isomorphic problems

The Problem Solver, No. 4 (Spring 1982) included a pair of puzzles, on different pages, with no suggestion that they might be connected. Indeed, I tried to disguise the fact. The first explained:

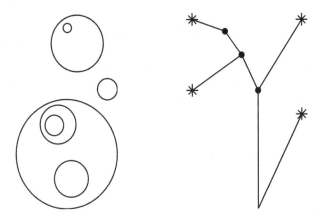

The bubbles match the trees (1987a Unit 61: F)

Here is one way to arrange 7 bubbles so that some of them are separate, some of them are inside each other, and some of them are inside a bubble ... which is inside another bubble ... and so on! In how many different ways like this can 7 bubbles be arranged?

The second puzzle read:

This BUSH has 4 flowers and 3 joints, making 7 joints and flowers in total. (The point where the root goes into the ground does not count in this problem.) How many bushes can you find with a total of 7 joints and flowers?

I deliberately included them both in the same issue, because I hoped that readers would realise that they are isomorphic — effectively identical. Not so. The next I heard of it, Don Steward had written an article titled 'Bubbles', for *Mathematics in School* (16-2), in which he described using the bubbles problem with his pupils, and the fascinating use they made of notations to describe different arrangements and hence to try to solve the problem.

This was followed by a letter from Roger Bray (*Mathematics in School*:16-4) in which he described using the same problem with his sixth formers. Bray and his colleague Richard Shephard then published a 61-page book, *From Bubbles to Trees: A Topic for Investigation* (Bray & Shephard 1987) that discussed the work of their pupils, including the work of one pupil (Rebecca) in full. In this discussion Bray and Shephard do discuss, 'Problem L.2: What is the number A_n of different rooted trees with (n + 1) vertices and n edges?' which they had previously concluded was equivalent to the bubbles problem (without being aware of the original 'tree' problem in *The Problem Solver*) which they then solved by following Cayley (1857). However, there is no emphasis in their discussion of the very idea of isomorphic problems.

This progress illustrates a curious feature of this pair of problems.* At first the isomorphism was not noticed, which suggests that teachers are not aware of ideas of structure in their teaching. In another situation, the two problems might have been naturally linked together by readers. When it is discussed at a relatively high level, the idea of isomorphism is not discussed either, presumably because it is, as it were, taken for granted.

---

* It suggests a second feature also: it was the bubbles problem which was a focus of interest. It appeared in the work of Steward and Bray and in SIGMA 1; the isomorphic *tree* problem appeared in an OCEA *Teachers Guide: Mathematics*, but nowhere else as far as I know.

I would like to suggest that for most secondary pupils the aspects of isomorphism and mapping are the most important aspects of these problems. There are many effective combinatorial counting problems, but relatively few examples of pairs of problems which appear to be completely different, but which turn out to be isomorphic. Yet such pairs, because they illustrate a deep and abstract mathematical concept in an active manner — it appears as a result of actively trying to solve problems — are a powerful illustration of those same concepts.

In particular, they illustrate one reason why ideas of structure are such a powerful tool for the mathematician. Why solve two problems when you can solve one?

Don Steward and Bray and Shephard in effect produce and discuss at length a considerable number of problems isomorphic to the bubbles problem, simple by trying to invent notations to suit the problem: each notation naturally leads to the question — 'How many expressions of this notational form are there?' The bubbles problem has naturally been transformed into an isomorphic problem about an algebraic expression or a type of diagram.

Rebecca, whose work Bray and Shephard reproduce, referred, they report, in discussion, to 'Bubblematically equivalent, as it were'. At her level, and at the level of pupils to whom the idea of transforming one problem into another is a novel concept — it was Don Steward's suggestion to his pupils that they try to represent arrangements of bubbles by notations — explicit possession of the idea of isomorphism, and the term for it, are a valuable accomplishment, which is at present denied them.

The bubbles problem duly turned up as an investigation in an exam by the University of London Examinations and Assessment Council, including the instruction to 'Create a symbolic system for representing the arrangements of bubbles. This system should be consistent', with a detailed marking schedule four pages long (ULEAC 1995).

Ambiguity: its rewards and difficulties

Mathematicians and the notations they use often exploit *structural parallels* to simplify notation. Experienced mathematicians are not misled because they appreciate the multiple interpretations involved. Pupils are easily misled, for example by the use of − 5 for the action, 'subtract 5', and the noun, 'negative 5'. A consequence is that so many pupils fail to either to understand or use negative numbers effectively, and so the topic become (yet another) source of demotivating failure.

Even more unfortunately, a confusing ambiguity appears at the start of arithmetic, with multiplication and division and then with *fractions*, which pupils start in *primary* school, but which also incorporate deep ambiguities. This poses no problem for experienced mathematicians because they appreciate the ambiguity, they know why it exists and why it doesn't matter: some pattern or structure lies behind the ambiguity and renders it not merely innocuous but powerful.

Mathematicians also know that when it is appropriate they can do operations 'by rote', without thinking about the meaning, but they can also interpret what they are doing when necessary. We want pupils to be able to do the same, to perform basic (game-like) moves smoothly and easily and correctly, but at the same time we want them to be able to pause, stop, and think about the meaning when it is necessary to do so. No wonder most pupils find mathematics hard!

The ambiguity of multiplication and division

Does 3 × 4 means three 4s or four 3s? It means either and it can mean both but this causes no confusion to most pupils for one of two reasons:

- Either they just 'know' that three 4s or four 3s are the same, 12, even if they can't explain why.
- Or they know this fact and can explain it with a picture.

So this simple ambiguity in the *notation* of multiplication does not really pose a problem, but note that if it did we could very simply write (for example) three 4s as 3(4) and four 3s as 4(3), emphasising that different numbers are inside the brackets and a different number outside, and inviting the question, 'Why are they equal?' more strongly than the 3 × 4 and 4 × 3 notation. (The bracket notation also looks forward helpfully to elementary algebra.)

Division incorporates the same ambiguity, but rather confusingly. Does 12 ÷ 4 mean 12 divided into 4 equal parts' or 'The number of 4s in 12?' These are distinct questions, albeit with the same answer, whose difference is highlighted if we insert a choice units: does 12 m ÷ 4 mean '12 m divided into four equal parts' or 'The number of 4s in 12 m?' It clearly means the former: by inserting a measurement of length, we have resolved the ambiguity, which is however resolved in the opposite direction if we ask: does 12 m ÷ 4 m mean '12 m divided into 4 equal parts' or 'The number of 4 m lengths in 12 m?'

Now the answer is the latter. No wonder that small children can find this confusing, especially when not enough emphasis is placed on the meaning of the question: children are often just too good at spotting patterns which work (or *seem* to work!) and sticking to them, even when they don't understand why they work. (Science teachers know this phenomenon. Pupils can bring plausible but wrong explanations to class, of which the teacher then has to tactfully but firmly disabuse them, while teaching them the sound explanation. This can be very difficult to do.)

The ambiguity of fractions is unfortunately even greater. Does ¾ mean three quarters added together, ¼ + ¼ + ¼? Or does it mean '3 divided into four equal parts?' Or does it mean 'How many 4s are there in 3?' Answer, it can mean any or all of these which is tough on primary pupils just starting on the topic!

(Textbooks can be confusing too: see the box overleaf.)

Pupils can be further confused by the reasons — or lack of them — for the different operations. Addition and subtraction require the fractions to be transformed to have the same denominator, and this approach also works for division:

$$2/5 \div 3/7 = 14/35 \div 15/35 \stackrel{\cdot}{=} 14 \div 15 = 14/15$$

Why can multiplication not be done using the same approach? Is it any wonder that pupils are puzzled?!

Textbooks on the ambiguity of fractions

How do textbooks cope with the ambiguity of fractions? About two years ago I checked 26 secondary maths series, some old, some new, but all in print, checking the chapters on fractions in the first book/volume/unit in each case. No less than 20 made no such reference to such ambiguity.

Modern Mathematics for Schools, Mathematics for Schools and *Shape and Number* explicitly recognised that 2/3, for example, is 1/3 + 1/3 and also two things divided equally between three people, but missed the third interpretation. *Longmans Mathematics* recognises that 3/4 is both 1/4 + 1/4 + 1/4 and also 3 divided by 4, and *Westminster Mathematics* says, 'The fraction 4/5 means '4 divided by 5' and its value is equal to 4 ÷ 5', but neither text recognises that 4 ÷ 5 is itself ambiguous.

... I also checked *The Mathematics Curriculum: Number* (Schools Council 1978). It starts well (36): 'In working with fractions, it is important to develop children's awareness of several related viewpoints. We exemplify three of these by the following simple questions ... ' But the first two are more or less equivalent from my present point of view, and later it says: 'The ideas to be assimilated seem to be: (i) 1/3 is the same as 1 ÷ 3 and means what we get when we divide a unit ... into 3 equal parts. (ii) 2/3 is 2 × 1/3 and 2 ÷ 3 and can be thought of either as two 'thirds' or as a third of 2.' Again missing the ambiguity of 2 ÷ 3.

Modern World Mathematics on the other hand makes no distinction explicitly but notes while introducing division of fractions, that 20 ÷ 4 asks the question, 'How many 4s in 20?' It does indeed (though that is not the only question it can be interpreted as asking,) just as 2 ÷ 3 asks 'How many 3s in 2?' and 10 ÷ 3 asks, 'How many 3s in 10?'

The rule for those few textbook authors who do not ignore the distinction entirely seems to be that, 'if a < b, then a ÷ b means a divided into b equal parts, but if a > b, then a ÷ b means how many bs in a.'

No wonder pupils are confused!

(1982:32–33)

11

Mathematics Applied to Science

It might seem obvious that the hard sciences which most pupils study in parallel with mathematics should be a major source of motivation for maths students. Amazingly, this is not the case. As we have seen, maths textbooks and syllabuses have little to say about applications in general and almost nothing about science in particular. Teachers and educators in these subjects seldom talk to each other and professional subject conferences are notable for the lack of cross-fertilisation.

This absence reflects back on the sciences also. Land described this reaction of his biology students whenever he introduced quantitative concepts: 'A sort of shudder runs through the class' (Land 1963:58).

Nearly 50 years before that comment, in 1919, the Mathematical Association Report recorded this evidence from 'A former head of a mathematics department':

> Now that we have reached a stage when elementary Physics and chemistry form an essential part of our school curriculum there does not appear to be any fundamental reason why the relationship of Physics at least, to Mathematics in our schools should not be of the same nature as that of Geometry or any branch of Applied Mathematics.
>
> (Mathematical Association 1919:25)

He is going too far: physics is more than applied mathematics. However, we can certainly agree that physics should be closer to mathematics than it is. The same report suggested that,

> It is a sound pedagogic maxim that the pupil should first of all examine the simple natural phenomena around him, and test and discover the mathematical laws that they obey.

(*Ibid.*:20)

This strongly suggests the possibility of linking school mathematics to elementary science.

The advantages and benefits of *mathematical laboratories* were a popular theme about that time but somehow teachers always found it easier after a few years to abandon such rich and ambitious projects and return to book work. The following quotation from *The Correlation of the Teaching of Mathematics and Science* is by John Perry in his report to the Mathematical Association conference describing the kind of experimental work that is found in many primary schools today, but not in secondary mathematics classes (Perry 1909:12):

> To teach him the use of decimals and to educate his hand and eye and judgement, you allow him to measure things. But how do you do it? Often in the most uninteresting way! He is made to measure a certain length, to weigh an object, etc., and his answers are compared with the real length, weight, etc. Contrast this with the following exercise. He is given a block of iron and he measures its length, 2.27 inches, breadth, 2.63 inches, thickness, 1.95 inches. He finds its volume to be 16.77 cubic inches ... He is given a cube 1 inch in edge of the same kind of iron. He takes it to the scales, and finds it weighs 0.26 pounds, so he computes the weight of his block to be 4.36 lbs. He now goes and weighs it, and is delighted. Do you see where the difference comes in, and how interesting it is to find his computation agreeing with reality?

Michel Helfgott has argued for this conclusion:

> Mathematics and physics have been closely intertwined since ancient times ... From a pedagogical perspective, an integrated approach to teaching, in which mathematics and science interact with each other, is a viable and desirable option. Also, a genetic approach provides an adequate historical framework that can enhance the learning process.

(Helfgott 2004:Abstract)

This also supports the recent emphasis on the value of bringing history of mathematics into the classroom.

The modern hard sciences are profoundly mathematical, extremely fascinating to many, and extremely beautiful. That means for us and our pupils, of course, that the phenomenon of the natural world which are explained by mathematical theories are not so much beautiful to look at (though they may be) but surprising and mysterious, often based on simple (and very mathematical) structures which we can observe, and explicable by simple mathematical relationships. Science teachers might object, I suppose, that maths teachers are pinching some of their entertaining experiments but frankly, that won't wash because science has so many and if maths teachers use a few of the simpler (and cleaner!) ones, for example from mechanics, science teachers still have plenty left. Moreover, science teachers should be pleased if the boundaries between the subjects are broken down, or at least made porous, because that development would be in their interests too.

The fact that mathematics is so successful in explaining scientific phenomenon is deeply mysterious. The Nobel Prize-winning physicist Eugene Wigner once argued that,

> the enormous usefulness of mathematics in the physical sciences is something bordering on the mysterious, and that there is no rational explanation for it … The miracle of appropriateness of the language of mathematics for the formulation of the laws of physics is a wonderful gift which we neither understand nor deserve.
>
> (Wigner 1960)

Pupils cannot appreciate science at Wigner's level but they can feel surprised that — for example — very simply mathematical formulae can fit scientific phenomenon and be used to *make predictions*, an enjoyable *activity*. We have already considered the value of presenting Pythagoras as a means of *making predictions*, but the principle is general. The very idea of making a prediction about the future, of 'anticipating nature', is highly motivating and prediction itself is a highly motivating *activity*, related, obviously, to challenge, mystery, and other aesthetic qualities.

I am happy to stick my neck out and predict that in years to come the current failure to link secondary mathematics closely to secondary science will be seen as totally bafflingly, totally irrational, and one of the great errors of current mathematics (and science) teaching.

Prediction within mathematics

The use of the verb *to predict* adds a frisson of excitement to activities which if expressed in flatter, ahistorical and 'logical' terms might seem far less enticing. But this added excitement is not spurious, it is genuine: by adding (as it were) real time as a factor, we are emphasising that mathematical conclusions enable us to *do* things, to be *active*, to *achieve* something, plain facts which ahistorical and logical language hides. Very often, dynamic language is more motivating than static.

The mathematical fact that, 'If the sum of the digits of a number is divisible by three, then so is the number' can be reinterpreted as a means of prediction: 'You can predict which numbers are exactly divisible by three, by summing the digits of the number.'

Pythagoras's theorem, as we noted, can be presented *actively* as a means of prediction, as can the closely related conclusions about right-angled triangles and the basic trigonometrical ratios, which can be used to predict the distances or heights of inaccessible objects or points.

More generally, any theorem once it is proved or we are otherwise convinced of its correctness, can be used as a means of prediction and, of course, theorems are, by pure and applied mathematicians.

If the classroom is on the first floor, then the corner of the building opposite (let us suppose) will not be directly accessible but the distance to it can be predicted by either drawing a scale drawing of a triangle which uses the line joining two points on the windows of the classroom as one side, or by basic trigonometry. The distance can then be checked by sending two pupils downstairs with a tape measure to check, literally, on the ground.

When presenting such methods of finding the height or distance of an inaccessible object, the emphasis should be placed, of course, on the apparent impossibility of the task, not only to attract the attention of pupils and to increase their motivation, but to maximise the chance that they will remember the lesson: psychologists have long known that emotionally charged events are better remembered than the affectively flat.

If pupils are learning basic algebra which could be exploited *at their level of understanding* in scientific problems — but is not — then they will lose a powerful source of motivation and 'What's the point?' yet again becomes a powerful question.

Mechanics and easy classroom experiments

It is no accident that those examples involve formulae or the results of calculation and that many scientific theories or models can be expressed by using formulae. This is the quintessential link between mathematics and the hard sciences which goes back to Archimedes and the origins of mechanics. Fortunately, there are many scientific phenomenon that can be modelled very simply by basic mathematics even without algebra and the scientific background can provide a powerful motivation and a brilliant answer to the question, 'What's the point of this?'

In particular, elementary mechanics provides many puzzles which are simple to set up and also very clean. Pupils should, of course, do actual experiments as far as this is practicable. To quote *A Study of Mathematical Education* (Branford 1908:254) itself quoting the great scientist Ernst Mach:

> If we know the (mechanical) principles like those of the centre of gravity and of areas only in their abstract mathematical form, without having dealt with the palpable simple (experimental) facts which are at once their application and their source, we only half comprehend them, and shall scarcely recognise actual phenomena as examples of the theory.
>
> Purely experimental in the initial stages, the teaching of mechanics should throughout, even in the advanced stages (including the Honours Standard offered at the universities) make constant appeal to actual experiment carried on in laboratories, if the effect is to be lasting and a knowledge of Nature, as mechanical, to be something more than words ...
>
> (Ernst Mach, *The Science of Mechanics*)

Where will a triangle balance?

Given an actual physical triangle, where is its centre of gravity? The puzzle of finding the *balance point* is far more motivating if the triangle is large and cut from wood or metal so that it feels physically heavy. The weight and solidity of the triangle increase the motivation of pupils because they can be experienced directly.

Beams and balances

Place a long strip of wood with a weight of 2–3 kg tied to one end on a table so that 20 cm or so sticks out over the end of the table. How hard do you have to push down on that end to raise the weight? The force required is surprisingly large and so the effect on pupils who try to lift the weight is large also.

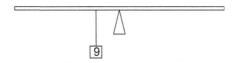

Where will 2 kg balance this weight?

This balance shows a 9 kg weight suspended 2 units from the centre. Where would a 2 kg weight have to be suspended to make the beam balance?

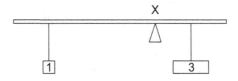

Where is point X? Why?

Here the weights have been placed already and the puzzle is to decide where they will balance. By experiment, the solution is the point X, dividing AB in the ratio 1:3. This answer seems convincing because it is so elegant and 'symmetrical' but how could we argue for it?

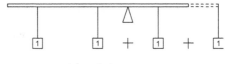

A heuristic argument

One plausible assumption is that we could replace the 3 unit weight with these three individual 1 unit weights. The centre of gravity then seems to be 'obviously' at X — but how justified was our 'plausible' assumption? Do actual physical weights really behave like that?

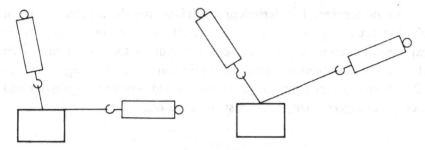

Experiments with spring balances

In these figures, the forces are more complex. What is the connection between the weights of the objects and the readings on each pair of spring balances (1987a:Unit 45; 1988c:Ch. 13)?

Rays of light

Mary, who is standing at S, wishes to walk to the river for a drink and then back to T, walking as short a distance as possible. To what point on the river bank should she walk?

—————————————————————————

RIVER

—————————————————————————

·T

S·

Mary walks to the river and back

This is the common version of the puzzle that Potts expressed in such abstract terms (page 110). The solution is both simple and elegant as we have seen, and involves reflection so it is appropriate that the same puzzle expressed as a problem about mirrors was solved by Heron of Alexandria (c. 75 CE) in his *Catoptrica*. He asked how a ray of light is reflected off a mirror and answered that it takes the shortest path and that its angles of incidence and reflection are equal.

It is no accident that Heron's problem is so important in the history of physics: according to modern science, all physical properties can be expressed in terms of extremes, so it is fortunate that, as we have seen already, puzzles about maxima and minima are so motivating, as well as often having a very practical value. Real-world problems in physics and mechanics suggest many puzzles, such as this one:

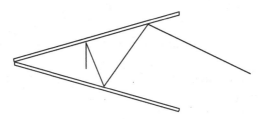

Where will the ray of light end up?

A ray of light is directed into the space between two mirror lines. What happens? Is the ray bound to emerge again? Could it retrace its inward path exactly? If it does eventually emerge, how long can its path be? And so on …

Parabola experiments

It is easy, though messy, to ink a marble, flick it upwards and across a large sheet of paper pinned to a rigid inclined board and so create an almost perfect parabola, allowing for the width of the ink trace:

Parabolas traced by an inked marble

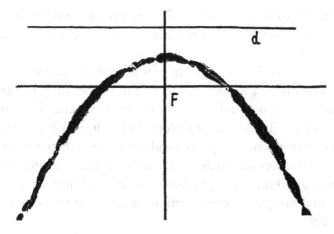

The axis and then the focus and directrix are identified

A vertical axis of symmetry can now be added with a ruler. If the ruler is then placed horizontally across the parabola so that the width cut off by the parabola is four times the distance of the ruler edge below the vertex the ruler will pass through the focus and the directrix will be an equal distance above the vertex.

It is easy to check that the distances of sample points on the parabola from the focus and directrix are equal, allowing for experimental error such as the width of the ink trace. This is an extremely good model of the flight path of a parabola through the air: the force of gravity has, in effect, been reduced somewhat by the use of the sloping board but in exchange, the pressure against the board creates the visible parabola. A good swop!

Science, aesthetics and mathematics

> You may object that by speaking of simplicity and beauty I am introduc-
> ing aesthetic criteria of truth, and I frankly admit that I am strongly
> attracted by the simplicity and beauty of the mathematical schemes that
> nature presents to us. You must have felt this too ...

> (Heisenberg to Einstein: quoted in Osborne 1984)

All these examples of, in this case, mechanics and mathematics share this additional feature: they are elegant — pretty — beautiful — even cool — and their aesthetic quality is linked to simplicity, surprise, mystery and expectations met or broken and other features that we expect to find in the purest mathematics.

This should not be a surprise. Insofar as mathematical worlds can be explored scientifically, we expect science and mathematics to share aesthetic features, and both science and mathematics search for power and simplicity. This is just as true of biology and ecology as it is of the more traditional 'hard' sciences of physics and chemistry and their many applications in technology and engineering, not forgetting the computer.

In particular, thinking of pupils with no especial talent for either subject, beautiful images are everywhere in science and many of them are mathematical.

Part of the problem may be that, as we have already suggested, most of these features cannot be explained at a level that pupils can appreciate but that problem applies to the science teacher also: it is no reason for not exploiting the wonders of science as (aesthetic) inspiration while only *explaining* some of the simpler features, at levels that pupils can grasp — as we have done with simple balances and forces.

The surprise is not that there are so very many connections between mathematics and science — that is obvious — but that these connections are so seldom exploited in the mathematics classroom, to our loss.

12

Classrooms in Other Countries

Extrinsic and intrinsic motivation in other countries

This book is concerned with intrinsic motivation but extrinsic cannot be ignored. In some countries (and in England in the past) mathematics has been much more highly valued than it is here today, and pupils therefore have, or had, a strong motivation to succeed, whatever the style or quality of their teaching.

Eastern European and Oriental students may have just such extrinsic motivation due to their social–cultural setting (and so may some British students, for example some Indian and Chinese, and of course those students who have just moved here from Russia or Japan).

Many students taking the International Baccalaureate (IB) examination need a strong motivation because they have to take mathematics whether they like it or not.* Dropping mathematics is not an option. (The UK is almost the only country in Europe where mathematics is not compulsory for post-16 students. Less than 10% of UK students then studied maths at that age (Adrian Smith 2004) though the numbers are rising.)

However, it is plausible that students from these countries enjoy stronger intrinsic motivation also, because of their very different styles of classroom teaching.

---

*This is certainly not a recommendation of the IB which in many respects is very abstract (even containing remnants of the Modern Mathematics Movement) and is not suited to many pupils.

Keeping pupils together

In all these countries there is an emphasis on the class working together and on the teacher keeping the class moving forwards together, linked historically and psychologically, plausibly, to the lesser emphasis on individualism in those societies. We are much keener on classifying and grading pupils into sets or streams within the one school, often creating humiliation and damaging the self-respect of those pupils who fall through the grating. (Setting and streaming also fits Anglo–Saxon enthusiasm for counting and measuring and so separating children by scores, even as the government promotes a culture of targets which scores the teachers and schools in national league tables and threatens underachievers with Hit Squads to sort them out.)

Typically, supporters of setting and streaming claim that life is very competitive and pupils should get used to the fact as soon as possible. They never point out that life is also extremely cooperative and that pupils ought also to learn to cooperate.

Adrian Smith's report (2004:3.42) emphasised the importance of the 'core skills' of *working with others* and *communication*, both developed in classes where interaction is rich and complex, but which are often not encouraged in English classrooms.

(Recall Cockcroft's suggestion (1982:464) that 'pupils who have become disenchanted with mathematics as a result of lack of success over the years can present problems of control in the classroom which make it difficult to continue oral work for any length of time'.)

Accepting differences in ability

Pupils know that some footballers are better than others though millions of players enjoy footballer with no chance of playing in the Premiership, or even in the local league. Millions of people can enjoy singing or play a musical instrument, without playing in an orchestra or becoming a soloist.

J. P., whenever he left his desk for any reason, had to be prevented from stopping off at other pupil's desks on the way and pointing out the solutions to problems, etc. He was just very smart at mathematics, and all the other pupils knew this but it didn't put them off, any more than they stopped playing football because one player in the team was outstanding.

What is damaging, even humiliating, and certainly demotivating, is being 'put down', by being shuffled off into a 'lower stream' or 'bottom set' where, in practice if not in theory, the teachers expectations are far lower and the pupils' expectations sink to match them.

Differences in expectations: Continental and Oriental classrooms

HM Inspectorate has consistently reported in its national surveys and in many reports on individual schools that a weakness far too frequently apparent in the present system is under-expectation by teachers of what they pupils can achieve.

That was the damning conclusion of *The National Curriculum 5–16: a consultation document* (#8(ii)) published in 1987.[†] Teachers on the Continent and in the Orient tend to have a less individualistic approach than the British. They emphasise the group and its success much more and the individual less. Therefore great efforts are made to keep the class together as they move forwards, and class discussion is emphasised. They also have higher expectations and so can push children harder.

Ability versus effort

In France, for example, all students follow the same curriculum in mixed-ability classes in all subjects up to the age of sixteen. This is premised on a constitutional right of equal access to the curriculum whilst, more generally, there is widely-held Eastern belief that all are educable and that it is effort not ability that determines success.

(Andrews 2008)

[†] The same document suggested (*Ibid.*:#57) that the Teachers' Regulations require all maintained schools to have staffs of teachers suitable and sufficient in number for the purposes of securing the provisions of appropriate education. In other words, there must be enough qualified teachers. There weren't then and there aren't today.

This conclusion that 'it is effort not ability that determines success' is at the very least an oversimplification — but it does have a powerful side effect, in this sense: as long as the criterion of success is the pupil's own effort, then the pupil is not undermined cognitively by failure: the conclusion will not be drawn that the pupils is stupid, only that he or she should have tried harder. As soon as understanding is emphasised then the pupil's failure is a reflection on their intelligence: they failed because they are thick, which is exactly what millions of our pupils think, and their teachers too — which is why they are in 4D and not 4A.

Recently, the idea that a focus on effort is more productive than a focus on ability has started to appear in Anglo–American psychological literature. Thus under the heading, 'The Secret of Raising Smart Kids', Carol Dweck suggests to parents:

> Don't tell your kids that they are. More than three decades of research shows that a focus on effort — not on intelligence or ability — is key to success in school and in life.
>
> (Dweck 2007–2008:37)

France

English visitors early in the twentieth century described French maths teaching as,

> … (emphasising) the exact presentation of theorems by students (often in front of a class ready to pounce on logical omissions) and on a rapid coverage to provide a grasp of the structure and unity of the whole subject at the expense of the endless working of examples which characterised English mathematics teaching.'
>
> (McLean 1990:20–21)

French education has a reputation for being relatively formal and rigid, but it can be innovative. As Alain Bouvier, author of *La mystification mathématique* (1981), once put it, when the inspectors are away, the teachers will experiment. Here is his account of an attempt to introduce 'open problems' into French classrooms:

> Teachers tend to argue that problem solving cannot help students to increase their mathematical knowledge … How will the theory appear? The theorems? The main tools? It is not possible to teach topics through problem solving. For several years now, in Lyons, I have emphasised the following simple ideas:
>
> - To do mathematics is, mainly, to solve mathematical problems.
> - Mathematics, by itself, is attractive. It depends on the chosen problems and on the way they are shown to the students. To be attractive, mathematics does not need to be wrapped up in 'stupid stories' or applied to other fields (even if it is useful and pleasant to apply maths to other topics).
> - For students and pupils, we must choose 'open problems'[‡] i.e. problems without immediate solution, without an immediate method of attack, and which are difficult to classify in terms of 'algebra', 'geometry' … Such a problem … Can be really, open for any mathematician. The most important characteristic is that such problems demand that something be created for its solution …
>
> (Bouvier 1983:6–7)

Alain Bouvier then described how problems are presented to pupils which they have not seen before. The pupils were divided into groups, and attempted to solve them by their joint efforts. After a specified time, the groups came together and a speaker for each group presented the groups' solution, or attempted solution, to the whole class who were expected to criticise it, while the group defend themselves.

[‡]These are some of the 'open problems' (very different from most English investigations) published in *93 problèmes pour nos élèves* (May 1982). An enlarged version, *250 problèmes pour nos élèves* was published in 1983 (Artigue and Houdemont 2007:373). The proportion of the 93 problems which much resemble English-style investigations is smaller than in my sample.

French 'open problems'

1 For which integers n, is the rational number,
1 + 1/2 + 1/3 + ... + 1/n, itself an integer?

6 If ABCDEFGH is a cube (sic), find a triangle PQR of minimum perimeter whose vertices are on the sides AB, CG and EH respectively.

25 Squares are constructed on the outside of the parallelogram ABCD, on each of its sides, AB, BC, CD and DA. What can you say about the centres of these squares?

42 The quotient of two integers less than 1000 is 0.6786389, by calculator. What are the two numbers?

43 How can pentagons be classified?

48 Given two polygons of the same area, can you dissect one to obtain the other?

65 AC = 8, DB = 10
Find x.

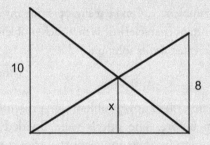

The crossing ladders puzzle

81 Given three concentric circles, construct an equilateral triangle with one vertex on each circle. What if the triangle is right-angled?

Hungarian classrooms

Among the goals of Hungarian mathematics education as recently as 1988 were

- 'Self-discipline and the sense of responsibility: accepting challenges and the ability to work hard over a long period of time.'

- 'Community spirit: cooperation, division of labour, assisting the less fortunate.'
- 'Moral courage: taking a stance, defending the opinion of an individual or that of a group even against the majority and against the status quo.'

(Nemelyi *et al.* 1988:107)

Since Hungary was still then a Communist country, the last point in particular may raise a smile although like all these points it is relevant to a particular style of teaching: the element of challenge, the high level of mathematical competence implicitly expected of the teacher plus the high expectations of the pupils, the importance of group work and cooperation, and the willingness to criticise and be criticised and to respond and defend yourself — features that English pupils and their teachers might find hard to match.

I will add that young Hungarian children are likely to be asked 'hard questions' by their grandparents and uncles and aunts (Julia Szendrei p.c.). These 'hard questions' are somewhat reminiscent of the traditional riddles that survived in our society into the Victorian era but have now died out, apart from some degenerate forms which appear in children's comics.

At a much more advanced level, the Hungarians are famous for their mathematical competitions for school children, and also, of course, for their many world-class mathematicians: two of the greatest were John von Neumann, the computer pioneer who also created modern game theory with Oscar Morgenstern, and the late Paul Erdös, the subject of the best-selling biography, *The Man who only loved Numbers*.

It must be noted, however, that there is — inevitably — an element of ideology in the Hungarian experience, rooted in Hungarian culture and history and modal personality and like all ideology both enabling and disabling. On the one hand, there is a very real attempt to keep all pupils in each class moving forward together, based on an ethos of high expectations and hard work. On the other hand, this ethos clearly fails for a proportion of pupils, and in addition, a cohort of pupils are early separated and sent to vocational schools. Therefore, there is no single, monolithic, Hungarian 'success story' that other countries can simply copy and imitate.

In Hungarian eyes, teachers in this country are unwilling to push children very hard. Two Hungarian visitors to this country wrote that,

we formed the opinion that in an English school the underlying philosophy of teaching is quite different from ours.... Not to hurt the 'self image'
of the children is more important than to force them to achieve better
results.

(Hatch 1994)

This is related to the problem of low expectations which is related in turn
to an English willingness to set and stream at the first opportunity on the
assumption that pupils of different abilities cannot work together (though
ironically top and bottom sets will be very mixed anyway).

Lack·of respect for authority among youngsters contributes to the
demands placed upon schools and teachers, who have to handle the problems created by society as well as teach their academic subjects: another
problem that is less serious on the Continent.

Japanese classrooms

Japanese practice also emphasises problem solving, and the experience of
the teachers, and cooperative working:

> Our analysis also suggests that lesson designs in Japan are also influenced
> by the occurrence of 'Lesson Studies' and by recent Japanese research
> into the learning and teaching of mathematics. For example, lesson study
> practiced by teachers in Japan for the last several decades, is one of the
> most common forms of professional development and involves teachers
> working in small teams collaboratively crafting lesson plans through a
> cycle of planning, teaching and reviewing.

(Yoshida 1999)

Through this process, Japanese teachers appear to have collaboratively
developed a view about good lessons of mathematics research that has
influenced how teachers structure lessons includes the work on the open-
ended approach in which the teacher gives the students a problem situation in which the solutions or answers are not necessary determined in
only one way (Sawada 1997:23). Considering all the influences described

above, in summary, Japanese teachers tend to structure maths lessons as follows (Stigler & Hiebert 1999:79–80):

1. Presenting the problem(s) for the day — designed to make students engage in mathematical thinking in challenging (or sometimes open-ended) situations — reviews of previous lessons are sometimes included before the problem(s).

2. Development of the problem(s) — students work on the problem(s) individually or in groups, discussion and presentations of solutions are often included, new (usually related) problems are sometimes introduced.

3. Highlighting and summarising the main point(s) — students' ideas are often used, and sometimes students are asked to explain their solutions; solutions are summarised by the teacher with the aim that the students grasp the main goals of the lessons.

(Jones, Fujita & Ding 2005)

The USA

The USA did poorly in the 1995 Third International Mathematics and Science Study (TIMSS; Beaton 1996.) These were the scores of American pupils compared to five other countries, at age 13:

Age 13	Int'nal	Eng	Fra	Hun	Jap	Sco	USA
Overall mean	49	47	51	54	67	44	48
Number	53	48	53	59	71	47	54
Geometry	49	49	58	52	70	46	44
Algebra	44	41	39	52	64	36	44
Data etc.	57	62	63	60	73	58	60
Measurement	45	43	43	49	63	40	36
Proportionality	40	38	38	41	55	34	38

(Andrews 2007:1)

The English pupils did not perform well either, beating France and Hungary in only one category out of six in each case, and Scottish students did even worse. And the Japanese beat everyone.

The Americans took their failure seriously. The detailed analysis by Schmidt *et al.*, *A Splintered Vision*, was written collaboratively with the U.S. National Research Centre for the Third International Mathematics and Science Study (Schmidt *et al.* 1997). One of their major conclusions, fitting *splintered* in their title, was that, 'U.S. textbooks included far more mathematics topics than typical internationally' (*Ibid.*:55). Matching conclusions were that,

> U.S. Mathematics and science teachers devote small amounts of instructional time to many topics. Even the few topics to which they devote more attention typically take up less than half of their instructional time ...

> Mathematics and science teachers instructional time allocations echo the inclusive, but breadth rather than depth, approach characteristic of our unfocused curricula ... (*Ibid.*:87)

> U.S. Textbooks clearly use variety and 'spiraling' presumably to help hold student attention. This is done at the price of focusing, of treating content in longer sequences. (*Ibid.*:102)

They add that,

> Only when a topic is considered as a focus or in a more holistic, extensive, or intensive way is it reasonable to include more demanding performance expectations for it. Only then are we likely to go beyond simple factual knowledge and routine procedures ... A side effect of a steady diet of fragmentation is that more complex demands seem often to be infinitely delayed. (*Ibid.*:102)

We may add that the Big Picture is also delayed for ever. U.S. teachers also use many more activities per lesson than the Japanese or German (*Ibid.*:102, 105).

Their analysis of U.S. textbooks and of teacher's instructional behaviour is fascinating, and so is their attempt to relate the pattern they perceive, the *Splintered Vision* of the title, to U.S. cultural features. In particular, they present four principles of 'incremental assembly' or mass production that they believe do 'underlie many and varied aspects of American life' and, they suggest, U.S. school education:

- Any complex object, process, or activity can be partitioned into component parts, subprocesses, or subactivities.

- All parts, subprocesses, or subactivities of the same type can be produced uniformly and interchangeably.

- All required parts, subprocesses, or subactivities when completed can be assembled and integrated to form the complete, complex object, process, or activity.

- Different required parts, subprocesses or subactivities can be completed by different workers, either simultaneously or at different times, to be available when needed. (*Ibid.*:95).

This 'fragmenting' approach is illustrated by the traditional emphasis on 'behavioral objectives' which had the unfortunate consequence that in order to specify any complex task required 'the construction of *hundreds* (sic) of specific objectives' as its supporters frankly admitted.

(As it happens, I worked briefly on programmed learning machines in 1963 at Ashford Tutor Machines. The linear programs were notable for the tedium of creating them matched only by the boredom of trying to work through them. The branching machines did have the potential to take into account a variety of student responses, though this could only be realised on a grand scale, in theory, with the arrival of the personal computer (1995b:Ch. 13).)

An obsession with 'behavioral objectives' naturally promotes testing, testing and yet more testing. This has long been a feature of U.S. education and has now arrived on British shores:

American teachers, who have experience of state-imposed curricula and testing from second grade (7- to 8-year-olds) onwards, reacted with horror to the prospects of the British introducing such tests. They emphasised the sense of failure such tests induce in some children: 'Test scores follow the children like an albatross', said one.

(Makins 1987 quoted in Greer & Mulhern 1989:297)

Today, Schmidt notes, the ideology of behaviourism has decayed, but its after-effects are still present in the U.S. educational system, not least in textbooks with 'fragmented content, limited expectations, and (requiring) short attention spans' (Schmidt *et al.* 1997:98). They conclude that, 'U.S. Teachers do what we ask them to do. We hand them fragmented, inclusive curricula — curricula 'a mile wide and an inch deep" and they teach them.

I have presented Schmidt's conclusions in some detail because for me they ring several bells:

- I strongly suspect that many English syllabuses and textbooks have similar faults.
- English pupils typically end up with a shallow understanding of a large number of topics — and shallow understanding is easily lost.
- The more-or-less rapid progression from one topic to the next is entirely consistent with Cockcroft's 25-year-old observation on the dominance of written work and the lack of discussion, especially as pupils get older (Cockroft 1982:464).
- They are also consistent with the large number of mathematics teachers in England with relatively low-level qualifications. The less confidence teachers have in delving into a topic, the greater their incentive to wrap up that chapter and start on the next.

Compare this exchange between Anne Watson on a visit to a Russian school, and some Russian kindergarten teachers still in training:

> I told them that one of the things that I was interested in was their attitude to mathematics, how they felt about mathematics and how they felt about teaching mathematics. And they looked very puzzled.
>
> 'Why are you asking this?' I said, 'Well, in our country there are many teachers of other subjects and of younger children who may be, or may have been, scared of maths.' This seemed to be quite a novel notion to them — the idea that you could be scared of maths; they didn't recognise it. And they said, quite sensibly 'But if they're scared of maths, how can they teach the children!'
>
> (Watson 1993:8)

Problem solving in the USA

Solving problems, especially problems you have never seen before, is almost the antithesis of instruction-by-objectives. The whole point of a problem is that it is not the rehearsal of what you have already learnt or even the assembly of several specific skills, but a more-or-less creative effort to 'think for yourself' or think with others in your group.

Progressive U.S. math educators have been promoting problem solving for decades but with limited success. The National Council of Teachers of Mathematics long ago announced that problem solving would be its 'theme for the 80s'. Its 1980 yearbook was titled *Problem Solving in School Mathematics*, and hardly any of 22 articles were about the traditional word problem.

The 1982 yearbook, *Mathematics in the Middle Grades*, kept up the pressure, with chapters on 'Problem Solving for all Students', 'Using learning centres in problem solving' and 'Wanted dead or alive: problem solving skills'.

Mathematics Teacher for October 1982 announced a 'Best Problem' contest and some of the entries were published in the 'Problems of the Month' department that started in January 1983.

A problem solving-based mathematics program for seventh grade pupils was described thus:

> Its goal is to help students become creators of knowledge, not just consumers of information ... Five understandings about the nature of mathematics are both the foundation and the desired outcome of the program ... mathematics is not the same as arithmetic; mathematics is not rigid and there can be many ways to solve a problem; mathematics is not magical or capricious, and we can find patterns or regularities to help us to solve problems; mathematics is all around us in familiar everyday things; and mathematics is creative, and we can enjoy asking and answering questions of our own.

Such approaches are potentially highly motivating, but the NCTM has had limited success in improving U.S. maths education. Among seventh and eleventh graders surveyed in the 1986 National Assessment of Educational

Progress, a majority of students agreed that math problems were solved by following a rule and roughly half believed that learning mathematics depended mostly on memorisation, presumably of the rules (Brown *et al.* 1988: quoted by Muis 2004:326).[§] Garofalo (1989) came to matching conclusions: math problems are solved by using the specific rules and procedures found in the textbook and taught by the teacher, and mathematics as a whole was viewed by students as highly fragmented (Garofalo 1989: Muis 2004:326).

Currently, U.S. math teachers are embroiled in the so-called 'Math Wars' in which the NCTM and the voices of reason are struggling against the back-to-basics brigades who accuse them of promoting Fuzzy Math, and even of being the New New Math, which is ironic because there is no connection at all between the two goals, though many professional mathematicians are now on the side of the back-to-basics just as they promoted the New Math in the 1960s.

English mathematics education

The distinctive feature of mathematical instruction in England is, that an appeal is there made rather to the memory than to the intelligence of the pupil.

(Demogeot & Montucci: *Report on English Education*)

The two Frenchmen drew their unflattering conclusion *circa* 1865: it may not have been the intention of English mathematics teachers then, it certainly is not today, but it is too often the practical result. The lack of focus on problem solving and cooperative work, on discussion and criticism, between confident teachers and confident pupils, is a marked feature of the English system when compared to the Continental or Oriental. English teachers are far too keen to get to the explanation, and so they end up by telling the pupils too much and allowing the pupils to develop too little.

[§] Compare Buxton in *Do You Panic About Maths?* (1981:115) who listed 'A collection of rules and facts to be remembered' as an example of a very common misperception of mathematics, which disturbed his experimental subjects: 'There was generally a wish to know why as well as how something worked.' One subject remarked, 'I don't like applying something when I don't know why it does what it does' (*Ibid.*:42).

The late Richard Skemp told of a primary school teacher who had astonishing success in getting her pupils through the 11-plus exam. The astonishment lay in her method: she just got the children to talk, talk and then talk some more, to her and each other, asking questions, answering questions, arguing and disputing, agreeing and disagreeing. At the time it seemed astonishing that no one had bothered to film her to show her class in action. In retrospect the lack of interest seems predictable.

The tendency to underemphasise group work and class discussion is, plausibly, linked in practice to the tendency to separate pupils too readily into sets and streams, lowering the motivation of the majority. We may recall Cockcroft's conclusion already quoted, that,

> ... pupils who have become disenchanted with mathematics as a result of lack of success over the years can present problems of control in the classroom which make it difficult to continue oral work for any length of time. However, lack of discussion almost certainly leads to further failure and so the problem is compounded. (Cockcroft 1982:464)

On the other hand, there are many classes where pupils would welcome more interaction and are perfectly capable of conducting themselves responsibly, but are not allowed to do so:

> We cannot abdicate from our responsibility to ensure that material will be taught in a suitable style. Many teachers do not allow argument and debate.
>
> (Swan 1989)

Swann adds that many teachers 'corrupted and distorted' the laudable intentions of the (School Mathematics Project) authors.

British and U.S. education have other features in common, not least low expectations:

> Sometimes one seems to note in the writings of experts in child development ... altogether too much maudlin sentimentality, which makes the child an exceedingly delicate and fragile being incapable of dealing seriously with real problems.
>
> (Brownell 1937:4)

The USA and UK also share the *fragmentation* that Schmidt referred to, which is linked in both countries to testing, testing and more testing against 'objective' objectives, at the expense of any deeper or broader understanding.

Russian lessons

The Russian case is especially illuminating. A key demand for a Russian is (or was — the times are changing) that he or she be 'together in spirit' with others. Two examples of this feature in practice: when a Russian is applauded by a group, after giving a speech for example, he joins in the applause; and a Russian Orthodox priest is not raised up in a pulpit but stands at the level of the congregation.

A classroom consequence is that it is much easier for pupils to argue and discuss and to make mistakes, without disturbing this taken-for-granted 'togetherness'. A pupil who makes a mistake at the blackboard or who answers a question wrongly does not feel humiliated because they always remain a part of the group.

In Russian pedagogical theory there is another consequence: the emphasis — now familiar from other examples — on effort and hard work rather than on natural talent. This emphasis promotes high expectations of all pupils: 'Work hard, and you too can understand algebra, geometry, and trigonometry!' We may consider such expectation to be unrealistic for many pupils, but it cannot be denied that there is a large element of self-fulfilling prophecy here. If pupils believe they can achieve, they very likely will, and their teachers will tend to become wound up in the same beneficent circle (1994b).

Paul Andrews on TIMSS

Paul Andrews has discussed the results of TIMSS (1996), which tested more than half a million students in more than 40 countries. Noting that Scotland and England did badly he proposed 'to review the literature describing mathematics classrooms around the world to identify the characteristics of, or factors which contribute to, effective teaching' (Andrews 2008:1).

He concluded that where mathematics is taught most successfully:

- Mathematics is acknowledged as difficult.
- Mathematics is viewed as a problem solving activity.
- Problems are chosen to exemplify mathematical generality.
- Teachers do not seek to reduce the complexity of mathematical ideas.
- Mathematics lessons are sequenced to emphasise coherence and continuity.
- Teachers do not shy from the vocabulary of mathematics.
- Students are expected to engage with proof and justification.
- The applications of mathematics are subordinated to the subject itself.
- Mathematical ideas are revisited constantly within the problems offered.
- Relatively little time is given to the practice of routine procedures or exercises.
- Teachers teach a class as a unit.
- Teachers take complete responsibility for the planning and delivery of their lessons.
- Teachers dominate the majority of the lesson with their talking, or managing the talk of others.
- Students are expected to operate in a public domain in which the preservation of their self-esteem is subordinated to their mathematical development.

(Andrews 2008:2–3)

I quote Andrews at length because a summary cannot do justice to the power of his conclusions. He then observes that this is emphatically *not* a picture of English mathematics teaching and quotes 'a year seven pupil's description of her experiences of a typical mathematics lesson':

Wait outside if she's not there. Come in if she's there. Sit down. And she tells us what we're gonna do. And she'll probably write up a few examples and notes on the board. Then we'll either get sheets handed out or she'll write up questions on the board. Not very often. We mainly get a text-book. We'll get pages. She'll write up what work to do, page numbers and exercise. And if you finish quick you may get an activity sheet. And that's about what happens.

(Clarke 1984:16)

Ouch. That account was written more than 20 years ago. Have such lessons disappeared? No.[5] Is English mathematics education today based on problem solving? No. Are students expected 'to engage with proof and justification'? No. Do our students spend many hours doing 'exercises and routine procedures'? Yes. Recall Cockcroft's reference to 'teaching of a kind which, instead of developing understanding, concentrates on the drilling of routines in order to answer examination questions. We therefore have a 'vicious circle' which is difficult to break' (Cockcroft 1982:445).

It would be naive in the extreme, as Paul Andrews emphasises, to think that we can simply import the mathematical and educational cultures of faraway countries as a short-cut to improving our own performance. We cannot. Nor is it obvious that all the features that he detects in the most successful countries ought to be transported: perhaps they need adapting to a different culture, or perhaps there are other routes to excellence in mathematics education — but if there are, then we have not yet discovered them.

His last point is especially telling in this respect. It is a repeated observation of English culture — as we have seen — that we are inclined to treat our little darlings too delicately and desist from putting too much pressure on them in case they crumble even as we simultaneously dump thousands of pupils into lower streams and sets where we know they do badly and they know that they can achieve little.

This state of affairs could be hinting at some deep and pathological feature of our societies: or perhaps we have here no more, or less, than the end result of another vicious circle already described in our recap of Cockcroft 464. There we have, in outline, a vicious circle which is closely analogous to the *vicious circles of decreasing expectation (and motivation)* which we have already discussed (page 24).

In that situation, the pupil's apparent failure is attributed entirely to the pupil's own inadequacies, and (with the best will in the world?!) the pupil is moved down a set, or stream, where the circular process is repeated.

---

[5] At this point I had better insert the usual qualification: there are of course many excellent teachers to whom this description is absolutely not applicable, and whose pupils, if selected for the next TIMSS survey, would be a credit to their teachers, themselves, their country. But there aren't enough of them, and too many pupils can identify with the girl quoted by Paul Andrews.

It may be that in the present case also, given the pupils' failure of behaviour (rather than cognitive failure) teachers simply respond by lowering the behavioural demands on the pupils: rather than expect them to listen attentively to the teacher and each other and to show a positive drive to join in matched by a willingness to cooperate with others, the teacher provides them with work which helps to pacify the classroom even as it *decreases the expectations* on the pupils and *decreases their motivation.*

The two vicious circles are precise analogues of each other, the one focused on cognitive achievement, the other on behavioural. They are both disastrous for our pupils — and they are pretty damaging for the teachers too.

Why do they appear in English and Scottish maths classes and not, say, Japanese? That is no doubt a cultural question and we cannot change our culture 'just like that'. We can however try to understand our own culture, and so adapt our teaching and our pupils' learning in a beneficial direction.

13

Rich Complexity, False Simplicity

Concepts can never be presented to me merely, they must be knitted into the structure of my being, and this can only be done through my own activity.

(Field 1950)

A superficial cognitive simplicity is purchased only at the price of affective impoverishment.

(1992)

Rich complexity — provided it can be grasped — is much more attractive than a false simplicity which is claimed to be 'easy' (and maybe 'basic' too) but which can only be appreciated as such by a sophisticated taste. Mathematics that is too simple or falsely simple *especially invites* the response, 'What's the point, miss?'

False simplicity is strongly demotivating. Rich complexity is highly motivating. Why are puzzles so popular? They are in one sense extremely 'hard' and 'difficult' and threatening — rather than easy and simple. So why do so many pupils enjoy tackling puzzles? Because they are full of meaning — and the pupil can put meaning into them — and if and when you solve a puzzle *you* get the credit.

False simplicity

'(traditional teaching) is a perfect example of ... 'false simplicity'. Superficially it may seem entirely rational to start with the very simplest exercises and progress steadily through more and more complex material to the final goal.

It is not rational because it assumes and implies a theory of children's learning which is false, and a theory of what it means to understand mathematics which is also mistaken.

It ... assumes that understanding of mathematics consists of understanding ... sequences of steps. This is also false. Any mathematicians' or any teacher's understanding of mathematics is far more complex and subtle than this, involving a multitude of nuances, connections, associations, images, subtleties of meaning, even feelings, which a straightforward explanation will neither evoke nor develop.

Consequently the apparent simplicity of the traditional approach ... is an illusion.

Moreover, by denying pupils any personal challenge beyond copying the teacher's explanation, it effectively strips the work of most of its emotional meaning. Whatever cognitive simplicity may be achieved ... is more than balanced by the affective vacuity of the material.'

(1987a:21–22)

Percy Nunn writing in *The Mathematical Gazette* for January 1912 argued that,

The practice of withholding practical applications until the boy has mastered his method in the classroom neglects the motive power that lies in the practical situation. It is based on the school master's favourite fallacy that 'you can't do a thing 'till you know how to do it'. This fallacy reverses the natural order in genuine intellectual activity, by placing the formal before the heuristic phase.

He then gave this example:

A too familiar instance has just been give me by a student as a piece of his personal experience when a schoolboy. His class in woodwork were taught a new joint which was subsequently to be used in making an

Oxford frame. All did the work so imperfectly that they were not allowed to proceed to the frame.

A second attempt produced equally unsatisfactory results. In despair the teacher gave them the frame to make — and almost every boy turned out a creditable article!

(Nunn 1912:215 footnote)

We might say that the teacher finally gave up his attempts to *instruct* his pupils and gave them a *puzzle* instead, and they solved it. Compare research in recent years which shows that people can very often solve arithmetical problems in, for example, the local market (or perhaps at the supermarket checkout) — in real life and in real time! — which they cannot solve when they are presented as out-of-context pencil and paper exercises in the classroom. Instruction has an important place but it cannot be a substitute for the students' own activity and the students' own creation on meaning.

The result of 'instruction' is often what I refer to as false simplicity: instruction may look easy but it isn't, it may seem simple but is actually a dessicated simplicity that minimises motivation, which is maximised when the situation is reversed, when more time is spent exploring and developing concepts through the pupils' own activity. That, of course, has always been the message of progressive theories of education.

Instruction also creates the possibility — and often the *probability* — of failure, by being too quick and hard for pupils to follow fast enough to really understand what is being 'taught'.

Rich topics and intuition

Pupils subjected to falsely simple programmes develop little intuition or none at all. To develop rich intuitions you need rich experience and this takes time, of course. If stronger pupils short-circuit this process and appear to gain intuition quickly, that is only because in one hour they get ten hours of 'average' experience. Returning to the metaphor of the *miniature world* of the problem, pupils need to explore these miniature worlds, to examine their features, and to become *familiar* with them.

I referred earlier to professional mathematicians who rather than simply read and study a paper by a colleague, choose to recreate it as far as possible by their own efforts. This also takes time, but the result is richer understanding and it is precisely this richer and deeper and more intuitive understanding that the professional seeks and which we should desire for our pupils.

Lasting buildings are constructed on solid foundations, strong walls and secure floors. Many pupils' mathematical understanding is built on insecure foundations with weak walls and damaged superstructure and the result is unfit for mental habitation.

Morality and teaching

We return finally to the problem of our perception of our pupils and our expectations of them. Compare these two quotations:

> (our less able children) are the children who speak Bernstein's 'restricted code', who have bodies and spontaneous impulses but little in the way of minds.
>
> (R.S. Peters 1980)

> Each of the young people who spent part of their youth under the guidance of Anton Makarenko was approached by this great teacher as an individual rich in vision, a potential creator of material and cultural wealth, capable of moral integrity and happiness.

Spot the difference! The first is from the philosopher of education, R.S. Peters, writing in the *Times Educational Supplement* (14.11.80). The second was written about the great Russian educator Makarenko, whose own vision still inspires thousands of teachers (Kumarin 1976: Preface).

Respect versus humiliation

When pupils are given syllabuses that have not been designed for them; when they are tested, tested, and then tested again on material that is unsuitable; when they are then blamed for not achieving levels which on any rational judgement they cannot be expected to achieve; when the *faults of the system* are loaded upon *their* shoulders; then they are not being respected and their relative failure is indeed an indictment of the system, not of them personally. Paul Andrews makes the point that lack of self-esteem is no problem in countries like Hungary and Japan: it is a problem here, and a very grave problem for many working class children rather than for middle class pupils, many of whom will do anything the teacher tells them, or asks them, and never once think of asking, 'What's the point of this?'

For many pupils, school mathematics is profoundly humiliating and contributes to their general — and justified — dissatisfaction with school.

Humiliation, putting down, demeaning, can have a far greater effect, in depth and in longevity, than praise: slights are recalled for far longer than plaudits. Mathematics is not the only subject to have this effect on its

Respect

'I believe that children are not respected, by what they are taught, not the system within which they are taught, and therefore in effect though not necessarily in intention, by their very teachers.

If an educational system were designed from scratch to minimise the achievement and maximise the difficulties of most children, the present system, which pays almost no attention to the variety of the children and presents its various subjects with almost no structure through time and with no clear and understandable aims, the present system, I say, would be a prize winning entry and the proposer if honest would sign himself Procrustes.

If a large number of average adults were similarly incarcerated for several years at a time, given no justification for what they were being forced to learn, punished for failing to reach the standards set by their teachers, and finally examined and many of them told that they had failed or only scored grade D or E (on subjects they did not want to study in the first place), I wager the adults would seem as incompetent, and as stupid and as unwilling and as unsuccessful as our schoolchildren.

Of course, the experiment could not be carried out. The greater self-awareness and age of adults would guarantee that they would rebel ... When children sensibly rebel, they are treated as offenders.'

(1977:14)

failures, but it is plausibly the most serious case because failure at mathematics can be and often is so complete, final and inescapable, and long-drawn-out — as pupils face work they simply do not understand and as a result of their failure, sink, and sink, and then sink again. Not only maths teachers need to find strong motivation for maths pupils, society needs to do so as well! The consequences of humiliating failure are too often dissident and disaffected pupils whose failure in mathematics has results far beyond the mathematics classroom. It is our responsibility to make sure that this does not happen to our pupils.

C.T. Daltry

I propose to conclude with an account of a talk given by Cyril Daltry (1902–1981) as a companion piece to the passages from J.W.A. Young with which I started. Daltrey started teaching under the influence of Percy Nunn and ended up lecturing to mathematics graduates at the London Institute of Education. He was president of the Mathematical Association 1972–73 and his obituarist (Penfold 1981) summarised his view of maths education by quoting from his presidential address:

> I find myself reflecting that methods (HOW?) matter far more than contents (WHAT?)

Daltry knew very well that any topic can be taught by many different methods, and he took for granted that teachers would contrast and compare their favourite methods, always trying to find a better one — an approach more reminscent of accounts of Japanese schools today, than English.

In the same address he suggested that 'the profound tragedy of much mathematics teaching is that by it children become more stupid, more confused, more lacking in confidence in their own powers than they would be without it' (Daltry 1973:154).

(He also included a great joke: A girl pupil once remarked to a student teacher, 'I suppose if you study algebra for years and years you learn what *x* really is!' (*Ibid.*:157).)

Twenty years earlier, Daltry had spoken at the 1954 International Congress of Mathematicians in the section on 'Philosophy, history and education'. This is a part of what he said:

> In this talk I hope to persuade members of my audience whether they regard themselves as mathematicians rather than teachers, or the other way about, that these two aims (to create mathematics and to communicate mathematics) are one: that to communicate mathematics most effectively the teacher should show the mind of the creative mathematician at work.
>
> It is natural to assume that because the final form of a piece of mathematics is an ordered structure built up brick by brick that it must be communicated to human beings in the same logical fashion. The assumption ignores the nature of the learner. In particular his whole emotional response to mathematics and to his teacher. Everything we have discovered about the art of teaching suggests that the over-riding consideration is the development of a close personal relationship between pupil and teacher. For this, teacher and pupil must engage in a joint enterprise. This cannot arise, for non-mathematical pupils, in a situation in which the teacher has to impose mathematics from without. There is a permanent, inherent tension or conflict in the teaching of mathematics ... a tension between logical and psychological ...
>
> The prevailing attitude to mathematics in many children is that of fear, which inhibits and paralyses. The origin of this fear is loss of confidence ... We must arouse and maintain his (sic) interest, we must simplify and select our material. This process of communication is not easy for a creative mathematician because it is concerned with personal values ...
>
> Assuming that we agree on interesting a pupil in his learning it follows that he must be involved personally, emotionally, in his mathematics. The best way to do this is to pose a problem, a challenge that is just difficult enough for the pupil to think his way through. If the challenge arises from the pupil's own situation ... so much the better ... I should define a problem as 'a difficulty, *appreciated* by the pupil, and *awakening* in him a desire for its solution. Notice that the pupil's situation is comparable to that of a researcher in mathematics.
>
> This approach follows the historical, the research line of development in which the tools, techniques, processes and principles of mathematics are devised in response to the challenge of problems ... This approach

from problem to principle is probably already familiar to many of my audience ...

Daltry then went on to refer to creativity in mathematics and to quote Wertheimer (whose book *Productive Thinking* was published in 1945) and the *Gestalt* psychologists: 'I am concerned to give confidence and success through discovery, through problem solving using restructuring and seeking insight.'
Daltry then cautions:

> Teachers of mathematics need to guard against misleading analogies derived from comparisons between mathematical structures and the growth of knowledge in youth. For instance the view that mathematics is built up like a wall, brick by brick, each item firmly cemented to its neighbour, is a dangerous guide. I prefer the analogy of crystals forming and growing in a solution. Nor is the insistence on firm foundations necessary: as we study mathematics we deepen our insights into foundations ... And all teachers need to guard against forms of education that inhibit creative activity: to insist continually on the right beginning, the right method, the *only* way.

Not only is Daltry's account very powerful but, astonishingly, he even manages, *inter alia*, to anticipate, attack and undermine aspects of the Modern Mathematics Movement which had not then been created! And perhaps the investigations movement too, for he even remarks: 'For instance, generalisation in mathematics is a noble concept: as an aim in the classroom it is insidiously dangerous.' He did not elaborate.

When the abstract Modern Mathematics Movement finally arrived, Daltry was appalled. In the same section of the same congress was a paper by G. Kurepa, reporting for the Inquiry of the International Mathematical Instruction Commission (IMIC) on 'The role of mathematics and mathematicians at the present time'. As readers may guess from that title, it anticipated in spirit and in flavour many of the claims of the abstract Bourbaki strand of the Modern Mathematics Movement. The AMMM is dead and gone leaving only traces of its poison in the earth. Daltry, touch wood, is alive and well.

Appendix: Jennifer Kano's Letter

Dear Mr. Wells;

Thank you for contacting me about my solution. The funny thing is, I was thinking about that experience recently and realizing how significant an event it was in my life.

At the time I was a single, working parent that had gone back to school as a part-time night student. I took a math course in college because I had to in order to graduate. Much to my surprise, that course turned out to be one of the courses (other than those in my major) that I enjoyed the most. My math professor encouraged me to become a math major based on my very strong performance in that class. I told him I was bad at math, that I had hit the 'math wall' in high school as a junior after getting an 'A' in geometry as a sophomore, which I really loved. I really didn't get algebra 2 in spite of the fact that I passed with a 'B'. I never took another math class until I had to in college.

However, I always enjoyed solving puzzles, and your book in conjunction with that college class was to me, just a lot of puzzle solving and it was a joy to discover that it was also real math!

I am sorry to have to tell you that not only did I not become a math major, I never took another math course. I am an electronic graphic designer — website developer and web database applications programmer. My majors in college were graphic design and mass communication. But solving that problem has been significant to my life in non-math ways. It showed me how effective it can be to look at a problem with 'virgin eyes'.

To be open to all the possibilities even if they fly in the face of common knowledge. I think the reason I was able to come up with that solution is that I had no preconceived notions about it. I just went at it with a love of geometry and puzzle solving and the tools I had learned in high school geometry class.

Since then, whenever I come across any thorny dilemma in my life be it professional, or personal, I try to take the same approach. I back away, let go of what I 'know', and try to look at it with 'virgin eyes'.

It's amazing what presents itself when I do.

So thank you for your wonderful book. I thoroughly enjoyed all of it. What I learned through the experience may not have motivated me to take a greater interest in math, but I think it made me a better person and a better 'thinker'.

My solution has never been published. After I got your letter about it, I gave a copy of it to my math professor, who was as thrilled with it as I was, and then I filed it away and went back to being a mother and a student that eventually became a graphic designer.

I am really tickled that you are publishing my solution. Let me know how I can get a copy of your new book when it becomes available.

Sincerely yours,

Jennifer Kano

Author's Bibliography

(1977–1980), *Acid Rain: Studies of Meaning, Language and Change*, Rain Publications, Bristol. Eight issues published before the title changed to *Studies of Meaning, Language and Change*.

(1980–1988), *Studies of Meaning, Language and Change*, **9–23,** Rain Publications.). A journal of articles several of which were on mathematics education and the philosophy of mathematics.

(1977), Education: Ideals, Appreciation and Respect, *Acid Rain*, **2**.

(1979a), Problems, Games, Familiarity, *Acid Rain*, **5**.

(1979b), Maths and Morality, *Acid Rain*, **7**.

(1979c), *Teaching and Appreciation*, 17.2.79. Privately circulated.

(1980), Distributing Blame, *Acid Rain*, **8**.

(1980–1983), *The Problem Solver* (eight issues). A magazine of problems for school pupils, and their teachers, with accompanying Newsletters. The problems were presented with no hints or aids and later collected in *Can You Solve These?* (1982–1984).

(1981a), Tao of Teaching, *Studies of Meaning, Language and Change*, **12**.

(1981b), *ATM Supplement*, 24 March.

(1982), *Three Essays on the Teaching of Mathematics*, Rain Publications, Bristol.

(1982–1984), *Can You Solve These?* (Books 1, 2 and 3), Tarquin Publications.

(1983), Scared pupils, scared teachers: *Projectile*, **1**, 8–9.

(1985), Cockcroft, investigations and confusion, *Mathematics in School*, 14–1, 6–9.

(1986a), *Problem Solving and Investigations*, Rain Press, Bristol. 2nd corrected edition 1987, 3rd enlarged edition 1993.

(1986b), Problem solving in the Anglo-Saxon World. Book chapter contributed to *Didactique des Mathématique: le dire et le faire*, edited by Alain Bouvier, Cedic/ Nathan, Paris.

(1987a), *Mathematics through Problem Solving*, Blackwell. Re-edited and republished as *Problem Solving for National Curriculum Mathematics* (1990).

(1987b), Proof and reasoning: letter, *Mathematics School*, Nov. 1987.

(1987c), The language of 'problem' and 'investigation', *Studies of Meaning, Language and Change*, **18**.

(1987d), General Concepts, and Teaching, *Studies of Meaning, Language and Change*, **18**.

(1987e), False Simplicity and True Simplicity, *Studies of Meaning, Language and Change*, **18**.

(1988a), Epistemology of abstract games and mathematics, *Studies of Meaning, Language and Change*, **20–21**, Nov. 1987–March 1988.

(1988b), Epistemology of Mathematics, and teaching, *Studies of Meaning, Language and Change*, **22**.

(1988c), *Hidden Connections, Double Meanings*, Cambridge University Press.

(1988d), General concepts and mathematics teaching, *Mathematics in School*, November 1988.

(1988e), Which is the most beautiful? *The Mathematical Intelligencer*, **10–14**, 301.

(1988f), (With Shirley Clarke) What's the point of games in the mathematics classroom? *Investigator*, **12**.

(1989a/1990), *Engaging Mathematics I, II and III*, West Sussex Institute of Higher Education/University of Chichester, England. Now available free online at the STEM website at www.nationalstemcentre.org.uk.

(1989b), Why do mathematics?, *Mathematics Teaching*, 127, Association of Teachers of Mathematics, reprinted in *Teaching, Learning and Mathematics* (1994).

(1990a), *Problem Solving for National Curriculum Mathematics*, Blackwell, Republication of an edited version of (1987a).

(1990b), Are these the most beautiful? *The Mathematical Intelligencer*, **12–13**, 37–41.

(1991a), *Games as a metaphor for mathematics*, Discussion Group on the Philosophy of Mathematics, British Congress of Mathematics Education, convened by Paul Ernest: 13–16 July 1991.

(1991b), 'Problem solving in the Anglo-Saxon countries' in Bouvier (1991).

(1992), *Ways of Knowing: The Nature, Learning and Teaching of Mathematics*, Rain Press. Privately circulated.

(1993a), *Problem Solving and Investigations*, (3rd enlarged edition) Rain Press, Bristol.

(1993b), When is a problem solved?, Workshop on How Mathematicians Work, HMW Group, *Bulletin of the Institute of Mathematics and its Applications*, IMA, Jan.–Feb., 1993, **29**, 24–26.

(1994a), Presentation on cross-cultural aspects of mathematics education at International Congress of Mathematical Education (ICMI), University of Maryland, MD.

(1994b), Contribution in, Coming up to Russian expectations, Dick Tahta *et al.*, *Mathematics Teaching*, **146**, 33–36.

(1994c), Anxiety, insight and appreciation, *Mathematics Teaching*, **147**, 8–11.

(1995a), 'Investigation and the learning of mathematics', *Mathematics Teaching*, 150, reprinted in Bloomfield and Harries (1998), pp. 40–44.

(1995b), *You are a Mathematician: a Wise and Witty Introduction to the Joy of Numbers*, Penguin, London (and Wiley, New York, 1997).

(1995c), Defective views (on mathematical proof), *Mathematics in School*, May 1995, 24–25.

(1997), *The Penguin Book of Curious and Interesting Mathematics*, Penguin.

(2007), *Mathematics and Abstract Games: an Intimate Connection*, Rain Press (now republished as Wells (2012)).

(2008a), The Mysterious Vecten Figure, *Infinity*, 2008/2001, Tarquin Publications.

(2008b), Wells, D.G. and Schultz, D. A Surprise with Parallel Lines: An exploration that went wrong, then right, *The Mathematical Gazette*, **92**, 162–164. The Mathematical Association.

(2008c), Cut off too soon, Alas!, *Mathematics in School*, pp. 21–23.

(2010), *Philosophy and Abstract Games*, Rain Press.

(2012), *The Third Entity: a Philosophy of Abstract Games*, Rain Press.

(2012), *Games and Mathematics: Subtle Connections*, Cambridge University Press.

Bibliography

Akin, E. (1993), *The General Topology of Dynamical Systems*, American Mathematical Society, Rhode Island.

Alpern, D. (2008), *Factorization using the Elliptic Curve Method* @ http://www alpetron.com.ar/ECM/HTM.

American Psychological Association (1953), *Ethical Standards of Psychologists*, Washington, DC.

Andrews, P. (2001), Mathematics education and comparative studies: what can we learn from abroad?, *The Scottish Mathematics Council Journal*, 30: 56–59.

Andrews, W.S. (1908/2004), *Magic Squares and Cubes*, Dover, New York.

Arnold, M. (1910), *Reports on Elementary Schools 1852–1882*, HMSO, London.

Artigue, M. and Houdemont, C. (2007), Problem Solving in France: Didactic and Curricular Aspects, in *ZDM Mathematics Education* (2007), 39: 365–382.

Ashcraft, M.H. and Ridley, K.S. (2005), Math Anxiety and its Cognitive Consequences, in Campbell 2005: 315–327.

Association for the Reform of Geometrical Teaching, Conference, 17 January 1871, in *The Mathematical Gazette*, 1911, 6, 91, Mar., 1911, 1–2.

Atiyah, M. (1982), What is Geometry?, The 1982 Presidential Address to the Mathematical Association, *The Mathematical Gazette*, 66, 437, 179–184.

Atiyah, M. (1984), An interview with Michael Atiyah, *Mathematical Intelligencer*, 6–1.

ATM (1969), *Sixth Form Mathematics Bulletin*, 1; ATM.

ATM (1994), *Teaching, Learning and Mathematics*, ATM.

Banwell, C.S., Saunders, K.D. and Tahta, D.S. (1972), *Starting Points*, Tarquin Publications.

Baron, M. (1973), *Number in Elementary Mathematics: a Course Book for Teachers and Intending Teachers in Primary and Secondary Schools*, Hutchinson Educational.

Beaton, A.E. *et al.* (1996), *Mathematics in the Middle School Years: IEA's Third International Mathematics and Science Study (TIMSS)*, Boston College, Boston MA, USA.

Billingsley, H. (1570), *Euclid*, quoted in Oxford English Dictionary, entry 'problem'.

Birkhoff, G.D. and Beatley, R. (1959/2000), *Basic Geometry*, 3rd edition, AMS Chelsea Publishing, Providence RI, USA.

Bogomolny, A. (2007/2015), Cut-the-Knot @ maa.org/editorial/knot/Jan2001.html.

Borwein, J.M. and Bailey, D.H. (2004), *Experimental Mathematics: Plausible Reasoning in the 21st Century*, A.K. Peters, Natick, USA.

Bouvier, A. (1981), *La Mystification Mathématique*, Herman, Paris.

Bouvier, A. (1983), Letter from France, *Problem Solver*, **8**, 6–7.

Bouvier, A., ed. (1991), *Manual de Didactique Action*, Nathan, Paris.

Bowles, C. (c.1807), *The Self Instructor, or Young Man's Best Companion*, Henry Fisher, Liverpool, England.

Branford, B. (1908), *A Study of Mathematical Education*, Clarendon Press.

Bray, R. and Shephard, R. (1987), *From Bubbles to Trees: A Topic for Investigation*, Department of Mathematics, University of Essex.

Brown, C.A., Carpenter, T.P., Kouba, V.L., Linquist, M.M., Silver, E.A. and Swafford, J.O. (1988), Secondary School Results from the Fourth NAEP Mathematics Assessment, *Mathematics Teacher*, **81**: 337–347, 397.

Brown, W. (1991), The Mandelbrot Set is as complex as it could be, *New Scientist*, 28 September 1991.

Browne, C.E. (1906), The Psychology of the Simple Arithmetical Processes: A Study of certain Habits of Attention and Association, *American Journal of Psychology*, **17**, 1–37.

Brownell, W.A. (1937), The Revolution in Arithmetic, *The Arithmetic Teacher*, 1–1: 1–5.

Buxton, L. (1981), *Do you Panic about Maths?*, Heinemann.

Campbell, J.I.D., ed. (2005), *Handbook of Mathematical Cognition*, Psychology Press, New York and Hove, Sussex.

Carr, M., ed. (1995), *Mathematics and Motivation*, Hampton Press.

Cartwright, M. (1955), The Mathematical Mind, *Mathematical Spectrum*, 2–2: 37–45.

Cayley, A. (1857), On the Theory of Analytic Forms called Trees, *Philosophical Magazine*,13, 172–176.

Cayley, A. (1883/1896), Presidential Address to the British Association, *The Collected Mathematical Papers of Arthur Cayley*, Vol. II, Cambridge University Press.

Clarke, D.J. (1984), Secondary Mathematics Teaching: towards a Critical Appraisal of Current Practice, *Vinculum*, 21–24, 16–21.

Clarke, S. and Wells, D.G. (1988), What's the Point of Games in the Mathematics Classroom?, *Investigator*, 12, SMILE Project.

Cockcroft, W. H. (1982), The Cockcroft Report, *Mathematics Counts*, HMSO.

Coleridge, H.N.C., ed. (1905), *Specimens of the Table Talk of Samuel Taylor Coleridge*, John Grant, Edinburgh.

Coulson, C.A. (1969), On Liking Mathematics, *Mathematical Gazette*, 53, 385.

Coxeter, H.S.M., (1969), *Introduction to Geometry*, 2nd edition, John Wiley.

Cross, K., ed. (1984), *Readings in Mathematics Education: Sixth Form Mathematics*, Association of Teachers of Mathematics.

Daltry, C.T. (1957), *Proceedings of the International Congress of Mathematicians 1954*, 3, North-Holland Publishing Co., Amsterdam.

Daltry, C.T. (1973), Difficulties: A Voice from the Past, Presidential Address to the Mathematical Association at the Annual Conference, April 1973, *The Mathematical Gazette*, 57, 401: 153–160.

David G. and Tomei, C. (1989), The Problem of the Calissons, *American Mathematical Monthly*, 96–96.

Davis, P.J. (1995), The Rise, Fall, and possible Tranfiguration of Triangle Geometry: a Mini-history, *The American Mathematical Monthly*, 102–103, 204–214.

Dieudonné J. (1964), Recent Developments in Mathematics, *American Mathematical Monthly*, 71–73, 239–248.

Dweck, C.S. (2007–2008), The Secret to Raising Smart Kids, *Scientific American Mind*, Scientific American, 18–26, 36–43.

Dyson, Freeman J. (1983), Unfashionable Pursuits, *The Mathematical Intelligencer*, 5–13, 47–54.

Ebbinghaus, H. (1885), *Memory: a Contribution to Experimental Psychology*, Duncker and Humblot, Leipzig. English translation (1913) by Ruger and Bussenius, Teachers College, Columbia University Press, New York.

Edmonds. B. (1983a), *Projectile*, **1**, January 1983, Leicester.

Edmonds, B. (1983b), *Projectile*, **2**, June 1983, Leicester.

English, L. and Bussi, M. (2008), *Handbook of International Research in Mathematics Education*, Routledge.

Etten, H., van (1633), *Mathematical Recreations, or a Collection of Sundrie excellent Problems out of Ancient and Modern Philosophers, both Useful and Recreative*, London.

Feynman, R. (1964/2005), *Feynman Lectures on Physics*, Addison-Wesley.

Field, J. (1950), *On Not Being Able to Paint*, Heinemann, quoting Follett 1924: 20.

Fishbein, E. (1987), *Intuition in Science and Mathematics*, Reidel.

Fletcher, T.J., ed. (1964), *Some Lessons in Mathematics*, Cambridge University Press.

Fletcher, T.J. (1972), *Linear Algebra through its Applications*, van Nostrand.

Foerster, P. (2006), *Algebra 1*, 3rd edition, Prentice Hall.

Follett, M.P. (1924), *Creative Experience*, Longman, Green and Co., New York.

Frame, M.L. and Mandelbrot, B.B. (2002), 'Chapter 3', *Fractals, Graphics, and Mathematics Education*, Mathematical Association of America, Washington D.C..

Garofalo, J. (1989), Beliefs and their Influence on Mathematical Performance, *Mathematics Teacher*, **82**: 502–505.

Gibson, C.G. (2003), *Elementary Euclidean Geometry*, Cambridge University Press.

Godfrey, C. and Siddons, A.W. (1903), *Elementary Geometry: Practical and Theoretical*, Cambridge University Press.

Goulding, M. (2003), *Learning to teach Mathematics in the Secondary School*, David Fulton.

Greer, B. and Mulhern, G., eds. (1989), *New Directions in Mathematics Education*, Routledge.

Griffiths, H.B. and Howson, A.G. (1974), *Mathematics, Society and Curricula*, Cambridge University Press.

Groot, A., De (1965), *Thought and Choice in Chess*, Mouton, The Hague.

Haggarty, L. (2002), *Aspects of Teaching Secondary Mathematics: Perspectives on Practice*, Routledge.

Halmos, P. (1974/1998), *Naive Set Theory*, Springer.

Halmos, P.R. (1980), The Heart of Mathematics, *American Mathematical Monthly*, 87–97, 519–524.

Hamming, R. (1980), *Coding and Information Theory*, Prentice-Hall.

Hanna, G. (1996), The Ongoing Value of Proof, *PME XX*, **1**, Valencia, Spain.

Hanna, G. and Barbeau, E. (2008), *Proof in Mathematics* @ www.math.toronto.edu/barbeau/hannajoint.pdf.

Hardy, G.H. (1915), Prime Numbers, *British Association Reports 1915*, 350–354, reprinted in *Collected Papers of G.H. Hardy*, 1 (1967), Clarendon Press, Oxford.

Hardy, G.H. (1920), Some Famous Problems of the Theory of Numbers, *Collected Papers of G.H. Hardy*, 2, (1967), Clarendon Press, Oxford.

Hardy, G. H. (1941/1969), *A Mathematician's Apology*, Cambridge University Press.

Hardy. G.H. and Wright, E.M. (1938/1960), *The Theory of Numbers*, Clarendon Press, Oxford.

Hatch, G. (1994), (responses to Anne Watson's article), *Mathematics Teaching*, 146, 29.

Hatch, G. (1995), If not Investigations — what?, *Mathematics Teaching*, 151, 36–39. Reprinted in Bloomfield and Harries (1998), 45–48.

Healy, L. and Hoyles, C. (1998), *Justifying and Proving in School Mathematics: Technical Report on the Nationwide Survey*, London Institute of Education.

Helfgott, M. (2004), Two Examples from the Natural Sciences and their Relationship to the History and Pedagogy of Mathematics, *Mediterranean Journal for Research in Mathematics Education*, 3: 1–2, 147–166.

Howson, A.G., ed. (1973), *Developments in Mathematics Education*, Cambridge University Press.

Howson, G. and Wilson, B. (1987), *School Mathematics in the 1990s*, Cambridge University Press.

Hoyles, C., Morgan, C. and Woodhouse, G. (1999), *Rethinking the Mathematical Curriculum*, Falmer Press.

Hudson, L. (1966), *Contrary Imaginations*, Penguin.

Hudson, L. (1968), *Frames of Mind*, Penguin.

Huntley, H.E. (1970), *The Divine Proportion*, Dover.

Hutchon, K. (1982), Persistence, in Wells (1980–1983), *The Problem Solver, Newsletter 3/4*.

Hutton, C. (1840), *Recreations in Mathematics and Natural Philosophy, a New and Revised Edition by Edward Riddle*, Thomas Tegg, London.

Johnson, V.E. (1889), *The Uses and Triumphs of Mathematics*, Griffith Farran Okedon and Welsh.

Johnstone-Wilder, S. and Mason, J. (2005), *Developing Thinking in Geometry*, Paul Chapman Educational Publishing.

Jones, K., Fujita, T. and Ding, L. (2005), Teaching Geometrical Reasoning: Learning from Expert Teachers from China and Japan, in D. Hewitt and A. Noyes (eds), *Proceedings of the Sixth British Congress of Mathematics Education*, pp. 89–96.

Kac, M. (1959), Primes Play a Game of Chance, in *Statistical Independence in Probability, Analysis and Number Theory*, Carus Mathematical Monographs, Mathematical Association of America.

Kac, M. and Ulam, S. (1968), *Mathematics and Logic*, Praeger.

Küchmann, D. (1985), 'Transforming Billiards into Diagonals', *Mathematics in School*, 14–1, 48–52.

Kendall, M.G. (1961), *A Course in the Geometry of n Dimensions*, Griffin.

Kline, M. (1973), *Why Johnny Can't Add*, St. Martin's Press, New York.

Krull, W. (1987), Translated B.S. and W.C. Waterhouse, The Aesthetic Viewpoint in Mathematics, *The Mathematical Intelligencer*, pp. 9–10, 48–52.

Krutetskii, V.A. (1976), *The Psychology of Mathematical Abilties in Schoolchildren*, Chicago University Press.

Kumarin, U.V. (1976), *Anton Makarenko: His Life and Work in Education*, Progress Publishers, Moscow.

Lakatos, I. (1976), *Proofs and Refutations: the Logic of Mathematical Discovery*, ed. Worrall and Zahar, Cambridge University Press.

Land, F.W. (1963), *New Approaches to Mathematics Teaching*, Macmillan.

Lionnais, F., Le (2004), *Great Currents of Mathematical Thought*, Dover.

Loomis, E.S. (1902), *Original Investigation, or How to Attack an Exercise in Geometry*, Ginn and Co., Boston, USA.

Loomis, E.S. (1940/1972), *The Pythagorean Proposition*, National Council of Teachers, Washington.

Love, E. (1987), Review of Wells 1986, *Problem Solving and Investigations*, 1st edition, *Mathematics Teaching*, **119**.

Mach, E. (1883/1960), *The Science of Mechanics*, Open Court Publ., Chicago.

Mack, A. (n.d.), *A Deweyan Perspective on Aesthetic in Mathematics Education* @ http://people.exeter.ac.uk/PErnest/pome19/Mack%20-%20Dewey,%20Aesthetics.doc.

MacLane, S. (1983), The Health of Mathematics, *The Mathematical Intelligencer*, 5–14, 53–56.

Makins, V. (1987), Early Testing deepens Sense of Failure, Americans Warn, *Times Educational Supplement*, 31 July.

Mason, J. and Sutherland, R. (2002), *Key Aspects of Teaching Algebra in Schools*, School Curriculum and Assessment Authority.

Mason, J. (2004), *Fundamental Constructs in Mathematics Education*, Routledge/Falmer.

Mauldin, R.D., ed. (1981), *The Scottish Book: Mathematics from the Scottish Cafe*, Birkhauser.

McLean, M. (1990), *Britain and a Single Market Europe: Prospects for a Common School Curriculum*, Kogan Page.

Mertens, R. (1987), *Teaching Primary Maths*, Edward Arnold.

Midonick, H. (1968), *The Treasury of Mathematics*, Penguin.

Muis, K. (2004), Personal Epistemology and Mathematics: a Critical Review and Synthesis, *Review of Educational Research*, 74–3, 317–378.

Nemelyi, E.C., *et al.* (1988), *The Teaching of Mathematics in Hungary*, National Institute of Education, Budapest.

Neumann, J., von (1947), The Mathematician, *Works of the Mind*, 1–1: 180–196.

Newman, J.R. (1956), *The World of Mathematics*, George Allen and Unwin.

Newton, D.P. (1988), *Making Science Education Relevant*, Kogan Page.

Nunn, P. (1912), The Aim and Method of School Algebra II: Methods of Teaching, *The Mathematical Gazette*, 6, 214–219.

Osborne, H. (1984), 'Mathematical Beauty and Physical Science', *British Journal of Aesthetics*, 24–4, 291–300.

Ozanam, J. (1712), *Cursus Mathematicus: or, a Compleat Course of the Mathematicks*, trans. J.T. Desaguliers *et al.*, London.

Pendlebury, C. (1897), *Arithmetic*, George Bell and Sons.

Penfold, A. (1981), Obituary of Cyril Tetlow Daltry, *The Mathematical Gazette*, 65, 434, 275–276.

Penrose, R. (1974), The Role of Aesthetics in Pure and Applied Mathematical Research, *Bulletin of the Institute of Mathematics and its Applications*, 10, 266.

Perry, J. (1909), Report on Conference, *Mathematical Gazette*, 5, 77.

Perry, B. and Dockett, S. (2002), *Young Children's Access to Powerful Mathematical Ideas*, in English and Bussi (eds), 2008, 81–111.

Peters, R.S. (1980), *Times Educational Supplement*, 14.11.1980.

Piaget, J. (1970), *Science of Education and the Psychology of the Child*, Orion Press.

Plowden, B. (1967), *The Plowden Report*, Central Advisory Council for Education.

Poincaré, H. (1897), Sur les Rapports de d'Analyse pur et de la Physique Mathmatique, *Report of the International Congress of Mathematics*, Zurich.

Poincaré, H. (1930/2003), *Science and Method*, Dover.

Polya, G. (1945), *How to Solve It*, Princeton University Press, Princeton, New Jersey.

Polya, G. (1954), *Mathematics and Plausible Reasoning*, Princeton University Press, Princeton, New Jersey.

Polya, G. (1965), *Mathematical Discovery: on Understanding, Learning and Teaching Problem Solving*, John Wiley.

Pott, R. (1865), *Euclid's Elements of Geometry*, Longman.

Pritchard, C., ed. (2003), *The Changing Shape of Geometry*, Cambridge University Press.

Rodd, M. and Monaghan, J. (2002), School Mathematics and Mathematical Proof, in Haggarty (2002:Ch. 5).

Rota, G-C. (1988), *Advances in Mathematics*, 67–1.

Royal Society/JMC (2001), *Teaching and Learning Geometry 11–19, Report of a Royal Society, Joint Mathematical Council Working Group*, Royal Society/JMC.

Russell, B. (1907), The Study of Mathematics, *New Quarterly*, 1, 29–44.

Sage, A.H. (1903), Some Observations on the Teaching of Physics, *School Science*, 3–2, 67–80.

Sawada, T. (1997), Developing Lesson Plans, in Becker and Shimada (1997).

Schmidt, W.H., McKnight, C.C. and Raizen, S.A. (1997), *A Splintered Vision: An Investigation into U.S. Science and Mathematics Education*, Kluwer Academic.

Shapiro, H. N. (1983), *Introduction to the Theory of Numbers*, Wiley.

Smith, A. (2004a), *Making Mathematics Count*, HMSO.

Smith, A. (2004b), House of Commons, 19 April 2004, document HC: HC-197-viii, Q750.

Smith, E. (1996), *Mathematics, Computers and People: Individual and Social Perspectives*, in Ernest (1996).

SMP (1980), *A-level mathematics: the Report of the Stoke Rochford Conference*, School Mathematics Project.

Spens, W. *et al.* (1938), Secondary education with Special Reference to Grammar Schools and Technical High Schools, *The Spens Report*, HMSO.

Steen, L.A. and Selbach. J.A. (1970), *Counter-Examples in Topology*, Hold, Rinehart and Winston, New York.

Steward, D. (1987), Bubbles, *Mathematics in School*, 16–2: 42–43.

Stigler, J.W. and Hiebert, J. (1999), *The Teaching Gap: best ideas from the world's teachers for improving education in the classroom*, Free Press, New York.

Swann, M. (1989), Appendix to Greer (1989).

Sylvester, J. J. (1904), *The Collected Papers of James Joseph Sylvester*, Vol. II.

Thom, R. (1973), Modern Mathematics: does it Exist?, in *Developments in Mathematical Education*, ed. A.G. Howson, Cambridge University Press: 194–210.

Thurston, W. (1995), On Proof and Progress, in Borwein and Bailey (2004:98).

Turkle, S. and Papert, S. (1992), Epistemological Pluralism: Styles and Voices within the Computer Culture, *Humanistic Mathematics Network Journal*, 7, 48–68.

Voltaire (1764), *A Philosophical Dictionary.*

Wagon, S. (1986), The Evidence: where are the Zeros of n(s)?, *Mathematical Intelligencer*, 8–4, 58–61.

Watson, A. (1993), Russian Expectations, *Mathematics Teaching*, 145, 5–9.

Wells, D.G. and Schultz, D. (2008), A Surprise with Parallel Lines: An Exploration that went wrong, then right, *The Mathematical Gazette*, 92, 162–164.

Wertheimer, M., (1945), *Productive Thinking*, Harper.

Wheeler, D. (1970), The Role of the Teacher, *Mathematics Teaching*, 59, reprinted in Cross (1984).

Wigner, E.P. (1960), The Unreasonable Effectiveness of Mathematics in the Natural Sciences, *Communications in Pure and Applied Mathematics*, 13, 1–14.

Wiliam, D. (1993), Paradise postponed?, *Mathematics Teaching*, 144.

Yan, L. and Shiran, D. (1987), *A Concise History of Chinese Mathematics*, Clarendon Press, Oxford.

Yoshida, M. (1999), Lesson Study in Elementary School Mathematics in Japan: a case study, Paper presented at the American Educational Research Association 1999 Annual Meeting, Montreal, Canada.

Young, J.W.A (1907), *The Teaching of Mathematics in the Elementary and Secondary School*, Longman, Green and Co.

Zbrodoff, N.J. and Logan, G.D. (2005), What Everyone finds: The Problem-size Effect, in Campbell (2005:331–344).

Index